BRAVERMAN'S
INTRODUCTION TO

The Biochemistry of Foods

Joseph B. S. Braverman, 1895-1962.

BRAVERMAN'S

INTRODUCTION TO

The Biochemistry of Foods

Completely new edition

by

Z. BERK

Department of Food Engineering and Biotechnology

Technion-Israel Institute of Technology Haifa (Israel)

Elsevier Scientific Publishing Company

Amsterdam - Oxford - New York

1976

ELSEVIER SCIENTIFIC PUBLISHING COMPANY
335 Jan van Galenstraat
P.O. Box 211, Amsterdam, The Netherlands

Distributors for the United States and Canada:

ELSEVIER/NORTH-HOLLAND INC.
52, Vanderbilt Avenue
New York, N.Y. 10017

Library of Congress Cataloging in Publication Data

Braverman, Joseph B S 1895-1962.
Braverman's Introduction to the biochemistry of foods.

Bibliography: p.
Includes index.
1. Food--Composition. 2. Biological chemistry.
I. Berk, Zeki. II. Title. III. Title: Introduction to the biochemistry of foods.
TX531.B69 1976 641.1 76-20475
ISBN 0-444-41450-9

Printed in The Netherlands

PREFACE TO THE SECOND EDITION

The "Introduction to the Biochemistry of Foods" was the last work of Professor J.B.S. Braverman, who died on March 21st, 1962, just a few months before the publication of the book.

In the thirteen years which have elapsed since the publication of the First Edition, food biochemistry research has been extremely productive in new knowledge. This new information has been incorporated in many excellent texts on individual subjects. However, the number of general textbooks of a comprehensive nature concerned specifically with food biochemistry, designed for the food science or technology students at the undergraduate level, is still much less than the demand for this type of a text would seem to justify.

To fill this textbook gap and in compliance with the request of the Braverman family, this revised edition was developed.

The planning of the text is based on the definition of its subject matter, Food Biochemistry, as "the study of the composition of foods and of the reactions which lead to changes in their constitution and characteristics". Just as in the first edition, equal emphasis is given to the definition of foods as chemical systems and to the description of biochemical processes of importance in foods.

The didactic character of the book is retained. Extensive reference to original research papers is omitted, thus providing a readability and continuity considered essential for an introductory text of this type. Major reviews and primary information sources are listed at the end of each chapter. These are supplementary reading suggestions to provide the interested students with direction through which broader and more extensive knowledge can be acquired.

I wish to thank my colleagues at the Department of Food Engineering and Biotechnology, Technion, for their valuable criticism and ask their forgiveness for the many hours which could have been used to serve the Department better but were instead devoted to writing. Thanks are due to Mrs. Sarah Lewis and Miss

Rivka Schnurman, who typed the manuscript, and to Mrs. Miriam Rauchberger, who prepared the new drawings.

To the memory of J.B.S. Braverman, this effort is humbly dedicated.

Z. BERK.

CONTENTS

Selected bibliography at the end of each chapter.

CHAPTER 1

INTRODUCTION

Enter to grow in wisdom

HARVARD

The world in which we live is composed of two main classes of matter: the inanimate matter and the organized living matter. While the former, which comprises rocks, stones and minerals, is made of such patterns as atoms, molecules and crystals, in themselves well organized entities, it does not, by and large, change appreciably except perhaps owing to erosion, weathering or to cataclysm. On the other hand, living matter, which comprises microorganisms, plants, animals and human beings, while also built of atoms, molecules and crystals, is composed of extremely complex, highly organized patterns, which have the ability to grow, to support themselves and to reproduce. In addition, living matter is in a state of constant and rapid metabolic change.

Living matter should, in its turn, be divided into two general groups: the *autotropic* and the *heterotropic* organisms. Organisms belonging to the first group embody all green plants and a limited number of microorganisms. These are capable of growth and reproduction by preparing for themselves all the required complex materials from very simple inorganic compounds: carbon dioxide, water and a few minerals, utilizing sunlight to provide the energy necessary for the synthesis of organic matter.

The heterotropic organisms, on the other hand, are strongly servile towards the former group: they cannot subsist, grow or reproduce without *food*, which they have to acquire from the autotropic organisms or by devouring other members of their own group. In other words, organisms of the heterotropic group get the required energy for their subsistence only by decomposing foodstuffs which are ultimately derived from plants. This group embraces all the animals, including man, as well as most of the microorganisms. It represents a long chain of parasitism, beginning with man and ending with the smallest microorganisms, all hanger-ons with respect to green plants.

As mentioned above, the heterotropic organisms require their food to be supplied by others, and so man seeks to obtain his supplies from plants and animals. However, at a very early stage in his development, man was confronted with the problem of how to preserve his food from deterioration as well as from deleterious changes in color, taste, flavor, texture and in nutritional value. Through the ages, he learned this art of food preservation by trial and error and much of his knowledge for selecting suitable foods came to him by intuition.

The scientific study of foods is, however, a comparatively new field. One should probably attribute its beginning to Louis Pasteur. During the last 150 years much has been achieved in elucidating the nature of the chemical and biochemical changes occurring in foods. Yet our ability to define and understand our foods as physical, chemical and biological systems is still very limited. The reasons for the relatively late development and slow progress of food science may be attributed to the complexity of its subject matter.

Complexity

Most of our foods are extremely complex mixtures of many thousands of chemical species. Three groups of organic substances, *carbohydrates*, *lipids* and *proteins*, together with *water*, constitute the bulk of most foods and usually account for more than 99% of their mass. However, the balance is made up of hundreds of thousands of various compounds, some of them present only in concentrations of parts per million or less. Usually, it is to those "secondary" components that we owe the characteristic taste, odor and color of the foods. The vitamins, so important for their nutritional functions, are also present in minute quantities. On the other hand, some of the trace components may possess undesirable physiological properties, such as inhibitory and toxic effects. The development of new analytical tools is leading constantly to the discovery and definition of additional food constituents but our knowledge of the composition of our foods is still very far from satisfactory.

Chemical activity

Most food constituents possess chemically active groups in their molecules. They are capable of entering into complicated series of reactions with each other and with the media surrounding the food (air, water, packaging materials, processing equipment). Many of these reactions are accelerated by the presence of biological catalysts: the *enzymes.* During preparation and processing, foods are often exposed to conditions of heat, moisture and concentration which induce additional reactions. As a result of this reactivity and the constant interaction between the constituents, foods should be regarded as rapidly changing chemical systems. Pragmatically these changes may be classified as desirable or objectionable. Understanding foods as biochemical systems is therefore of utmost importance to the food technologist, whose task is, after all, to bring about the desirable changes while preventing, as much as possible, the undesirable ones.

Physical structure

The physical structure of foods determines an important factor of acceptability, namely the *texture.* It may also affect indirectly the chemical reactivity of a food. With very few exceptions, foods are not well-defined physical systems such as ideal solutions or crystalline solids. On the contrary, they are characterized by the presence of a multitude of interacting phases. The texture results primarily from the presence of macromolecular components and phases in colloidal dispersion.

We have defined "food biochemistry" as the study of the composition of foods and of the reactions which lead to changes in their constitution and characteristics (see Preface to the Second Edition). The following chapters describe the chemical nature of the principal constituents of foods as well as the biochemical processes in which these constituents are involved.

SELECTED BIBLIOGRAPHY TO CHAPTER 1

Baldwin, E., Dynamic Aspects of Biochemistry, 5th ed., Cambridge Univ. Press, 1967.

Bonner, J. and Varner, J.E., Plant Biochemistry, Academic Press, New York, 1965.

Chichester, C.O., Mrak, E.M. and Stewart, G.F., Advances in Food Research, Vol. 20, Academic Press, New York, 1948-1973.

Eskin, N.A.M., Henderson, H.M. and Townsend, R.D., Biochemistry of Foods, Academic Press, New York, 1971.

Hulme, A.C., The Biochemistry of Fruits and their Products, Vol. 1-2, Academic Press, New York, 1971.

Karlson, P., Introduction to Modern Biochemistry, 3rd ed., Academic Press, New York, 1971.

Kretovich, V.L. and Pijanowski, E. (Eds.) Biochemical Principles of the Food Industry, Proc. 5th International Congress of Biochemistry, Vol. VIII, PWN Polish Scientific Publishers, Warsaw, 1963.

Needham, J. (Ed.), The Chemistry of Life, University Press, Cambridge, 1970.

CHAPTER 2

PROTEINS, IN GENERAL

Of the first rank

BERZELIUS

2-1. OCCURRENCE

Proteins are very complex organic substances, present in all types of living matter: plants, animals and microorganisms. Their importance in life was recognized long ago by intuition. The name "*protein*" (from the Greek adjective πρωτειοζ, meaning most important, of the first rank) was suggested by Berzelius and first used by the Dutch chemist Mulder in 1838. In the animal body, proteins constitute the most important structural element, being the main component of muscles, skin, hair, connective tissue, etc. They occur in high concentration in seed cotyledons. Enzymes and many hormones are proteins. Thus, proteins are essential for life at all levels of organization.

All proteins contain, in addition to carbon, hydrogen and oxygen also nitrogen and often sulfur as well as phosphorus. The presence of nitrogen is an important characteristic, as it imparts to the proteins many of their peculiar properties.

2-2. THE PEPTIDE BOND

In spite of their complexity and immense diversity, all proteins have been found to consist of only about 20 structural units, the so-called "*amino acids*" (the so-called "*conjugated proteins*" contain additional groups such as sugars, lipids, phosphoric acid, nucleic acids and others).

All but two amino acids regularly found in proteins are of the general formula:

$$R-\underset{\displaystyle NH_2}{\underset{|}{CH}}-COOH$$

i.e. α-amino carboxylic acids, where both the carboxyl and the amino groups are attached to the α-carbon atom. In proteins the various amino acids are linked by *peptide bonds* (—CO—NH—), i.e. the carboxyl group of one amino acid is linked with the amino group of the second amino acid with elimination of H_2O, thus forming an amide of the second acid:

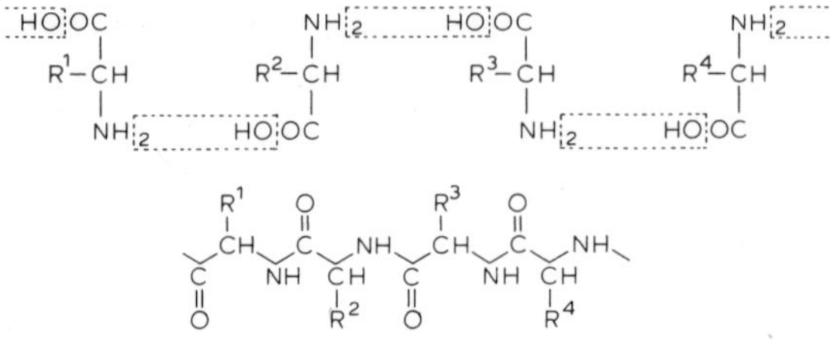

$$\begin{array}{cccc} HO|OC & NH|_2 \quad HO|OC & & NH|_2 \\ R^1-CH & R^2-CH & R^3-CH & R^4-CH \\ NH|_2 \quad HO|OC & & NH|_2 \quad HO|OC & \end{array}$$

$$-C(=O)-CH(R^1)-NH-C(=O)-CH(R^2)-NH-C(=O)-CH(R^3)-NH-C(=O)-CH(R^4)-NH-$$

the peptide bond

R^1, R^2 etc., indicate different amino acid residues.

2-3. AMINO ACIDS

When hydrolyzed by strong mineral acids or with the aid of certain enzymes, proteins can be completely decomposed into their component amino acids.

In the simplest amino acid, glycine, R is H. The following is a list of the most common proteinogenic amino acids:

(I) Aliphatic monoamino monocarboxylic amino acids:

Glycine $-CH_2\cdot(NH_2)\cdot COOH$

Alanine $-CH_3\cdot CH\cdot(NH_2)\cdot COOH$

Valine $-\ (CH_3)_2>CH\cdot CH\cdot(NH_2)\cdot COOH$

Leucine $-\ (CH_3)_2>CH\cdot CH_2\cdot CH\cdot(NH_2)\cdot COOH$

Isoleucine $-CH_3\cdot CH_2\cdot CH(CH_3)\cdot CH\cdot(NH_2)\cdot COOH$

Serine $-HO\cdot CH_2\cdot CH\cdot(NH_2)\cdot COOH$

Threonine $-CH_3\cdot CH(OH)\cdot CH\cdot(NH_2)\cdot COOH$

Proline $-$ (pyrrolidine ring, NH)–COOH

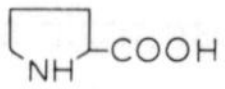

(II) Sulfur-containing amino acids:

Cysteine $-HS\cdot CH_2\cdot CH\cdot(NH_2)\cdot COOH$

Cystine $-$ $S\cdot CH_2\cdot CH\cdot(NH_2)\cdot COOH$
$|$
$S\cdot CH_2\cdot CH\cdot(NH_2)\cdot COOH$

Methionine $-CH_3\cdot S\cdot CH_2\cdot CH_2\cdot CH(NH_2)\cdot COOH$

(III) Monoamino dicarboxylic amino acids:

Aspartic acid – $HOOC \cdot CH_2 \cdot CH \cdot (NH_2) \cdot COOH$

Glutamic acid – $HOOC \cdot CH_2 \cdot CH_2 \cdot CH \cdot (NH_2) \cdot COOH$

(IV) Basic amino acids:

Lysine – $NH_2 \cdot CH_2(CH_2)_3 \cdot CH \cdot (NH_2) \cdot COOH$

Arginine – $(H_2N)(HN{=})C \cdot NH \cdot CH_2 \cdot CH_2 \cdot CH_2 \cdot CH(NH_2) \cdot COOH$

Histidine – $N{=}CH{-}NH$ ring; $CH{=}C \cdot CH_2 \cdot CH(NH_2) \cdot COOH$

(V) Aromatic amino acids:

Phenylalanine – $C_6H_5 \cdot CH_2 \cdot CH(NH_2) \cdot COOH$

Tyrosine – $HO \cdot C_6H_4 \cdot CH_2 \cdot CH(NH_2) \cdot COOH$

Tryptophan – C_6H_4 ring fused with $NH{-}CH{=}C \cdot CH_2 \cdot CH(NH_2) \cdot COOH$

Glutamine and asparagine, the amides of the respective amino acids as well as the hydroxy derivatives of lysine and proline, appear also in proteins, the last two mainly in connective tissues.

4-hydroxyproline (OH on the pyrrolidine ring, N–H, COOH)

5-hydroxylysine: $NH_2 \cdot CH_2 \cdot CH(OH) \cdot (CH_2)_2 \cdot CH(NH_2) \cdot COOH$

Except for glycine, all amino acids are optically active and in each of them the carbon atom, to which both the amino group and the carboxyl are attached, is the center of asymmetry. All amino acids forming natural proteins are of L-configuration. The D-forms can be synthesized, but only in exceptional cases are they found in biological material.

Due to the presence of both an amino group (a proton acceptor) and a carboxylic group (a proton donor), an aqueous solution of an amino acid should be regarded as containing the three following ionic species in dynamic equilibrium:

$$\underset{(a)}{R{-}CH(NH_3^+){-}COOH} \underset{1}{\rightleftharpoons} \underset{(b)}{R{-}CH(NH_3^+){-}COO^-} + H^+ \underset{2}{\rightleftharpoons} \underset{(c)}{R{-}CH(NH_2){-}COO^-} + 2H^+$$

Obviously, at higher concentration of H^+ ions (acid medium), the cationic form (a) is predominant, while the anionic form (c) is more stable in basic medium. The electrostatically neutral form (b) is known as a *zwitterion* or dipolar ion. Its net charge being zero, the zwitterion does not travel when subjected to an electrostatic field. At a specific value of pH, the amino acid will consist almost entirely of the zwitterion. The value of this pH, known as the *isoelectric point*, can be calculated from the dissociation constants of the amino and carboxyl groups. If we denote by K_1 the dissociation constant of the carboxyl group (reaction 1) and by K_2 that of the amino group (reaction 2), the isoelectric point pH_i of a mono-carboxyl-mono-amino acid will be found from the following relationship:

$$pH_i = \frac{1}{2}(pK_1 + pK_2)$$

where pK_1 and pK_2 are the negative logarithms of the respective dissociation constants K_1 and K_2. For glycine, $pK_1 = 2.34$ and $pK_2 = 9.60$ and $pH_i = 5.97$.

For most monocarboxyl-monoamino acids pK_1 and pK_2 do not differ much from the values given for glycine and the isoelectric points of those occur between 5.5 and 6.0. Dicarboxylic amino acids have their isoelectric point in the more acidic region (2.77 for aspartic, 3.22 for glutamic) while the opposite is true for amino acids containing more than one basic group (7.59 for histidine, 9.74 for lysine and 10.76 for arginine).

2.4. CLASSIFICATION OF PROTEINS

Complexity and diversity are the predominant characteristics of proteins. Rigorous classification of proteins is therefore difficult. It would require such a complicated system of definitions that the classification itself would become useless. A system of protein classification was suggested at the beginning of our century when information on protein structure was scanty. The criteria considered for classification were *solubility*, *coagulation* and *participation of residues other than amino acids (prosthetic groups)* in the composition. This classification is often meaningless as far as the more fundamental biochemical properties are concerned. Despite this and other inadequacies, the old nomenclature is still used and provides, at least, a certain degree of convenience. We shall describe this system of classification briefly.

First, we distinguish between "*simple proteins*" (the use of the adjective "simple" in connection with proteins is unfortunate) and "*conjugated*" or "*complex proteins*". Simple proteins are those which yield only amino acids on hydrolysis. Complex proteins contain nonprotein entities attached to the polypeptide chain.

(A) *Simple proteins* are further classified as follows:

(I) *Scleroproteins* are essentially insoluble, fibrous proteins with a relatively high degree of crystallinity. They are resistant to most enzymes and fulfill structural functions in the animal world. *Collagens* are the main binding agent in bones, cartilage, connective tissue and epidermis. *Keratins* are the structural element in hairs, nails, etc. The silk fiber contains *fibroin* and *sericin*.

(II) *Spheroproteins* or *globular proteins* have molecules of a more or less spherical shape. They are subdivided into five classes according to solubility:

(a) *Albumins*—Soluble in water and dilute salt solutions, they are precipitated by ammonium sulfate at near saturation concentrations. Examples: *ovalbumin* in egg white, *lactalbumins* in milk, the albumins of blood serum, etc.

(b) *Globulins*—are generally insoluble in water but soluble in dilute salt solutions. Examples: *Myosin* in muscle, *lactoglobulins* in milk, *glycinin* in soybeans, *arachin* and *conarachin* in peanuts, plasma globulins in blood, etc.

(c) *Glutelins*—insoluble in the above, but soluble in alkali and acids, viz. *wheat glutelin*, *oryzenin* in rice.

(d) *Prolamines*—soluble in 50-80% ethanol, viz. *wheat gliadin*, *zein* in maize.

(e) *Histones*—proteins of comparatively low molecular weights containing a large number of basic amino acids; they are soluble in water and in acids, viz. histones of fish spermatazoa, *globin* in blood.

(B) *Conjugated proteins* are usually named according to the nature of their prosthetic group, as follows:

Phosphoproteins contain phosphoric acid attached to the chain in ester linkage with the hydroxyl groups of serine and threonine, viz. the *caseins* of milk and *vitellins* of egg yolk.

Glycoproteins and *proteinpolysaccharides* contain carbohydrates. They differ in the length of the carbohydrate chains attached to the polypeptide moiety. In glycoproteins the carbohydrate consists of short oligosaccharides, while in protein polysaccharides the carbohydrate chains may contain hundreds of glycocyl residues. These two groups are important mainly as structural components in the animal body. They are widely distributed in connective tissues. Their presence in mucous secretions gave them the name of mucoproteins, by which those two groups were formerly designated.

Lipoproteins are proteins conjugated to lipids. However, the nature of the bond between the lipid and the protein in these substances is not always known. Since the lipid component can be often removed by extraction with solvents, covalent bonding between the moieties may be questionable. Lipoproteins of the blood plasma seem to fulfill physiological functions of vital importance.

Chromoproteins, in which the prosthetic group is a metalloporphyrin, such as chlorophyll or hemin (*in hemoglobin*). The group includes also the enzymes *peroxidase*, *catalase*, *cytochrome* and the *myoglobins* of muscle.

Nucleoproteins in which the prosthetic group, *nucleic acid* (Section 6-4-4), is in salt linkage with the protein.

2-5. PROPERTIES OF PROTEINS

2-5-1. *Amphoterism.* Although proteins are built only of amino acids linked together by peptide bonds, as shown before, they have specific properties of their own with the general properties of polymers of very large molecular weights. In the first place all of them are of a colloidal nature, and in addition they behave like dipolar ions, i.e. they exhibit amphoteric properties in similarity to amino acids. Since in proteins, the α-amino and the α-carboxyl groups of the amino acids are engaged in the peptide linkage, their amphoteric properties depend, by and large, on the ionizable groups in the amino acid side chains, such as the additional ϵ-amino group of lysine, the guanidino group of lysine, the γ-carboxyl of glutamic acid and the sulfhydryl group of cysteine, etc.

Electrostatic charges on the protein molecule determine to a large extent the forces within the chain, between chains and between the protein and the surrounding medium (polar solvents, other ions, etc.). Many of the external properties of proteins are therefore affected by the pH. Solubility, viscosity, solvation and optical rotation assume minimum values at the isoelectric pH. The adsorption of proteins by ion exchange materials depends on pH. For instance, at pH values below the isoelectric point the net charge on the protein is positive and the molecule is adsorbed more strongly on cation exchangers such as sodium carboxymethyl cellulose. This property permits separation and purification of proteins by *ion-exchange chromatography.* The mobility of protein molecules in an electrostatic field also depends on the pH. This property enabled Tiselius to work out the method of *electrophoresis* which has become one of the most valuable tools in protein research.

2-5-2. *Solubility.* The solubility of proteins as a basis for classification has been discussed above. Differences in solubility are often utilized in protein research for preparation, separation, purification and characterization purposes.

Differential precipitation (salting out) of proteins from their solutions by means of ammonium sulfate is a useful method of recovery and separation.

2-5-3. *Color reactions.* Several reagents give characteristic color reactions with proteins. In many cases the reactions are due to the presence of particular groups in the protein. A number of these reactions are also useful in the quantitative determination of proteins. A few are described below:

The xanthoprotein reaction. When heated with concentrated nitric acid, proteins yield a white precipitate which soon turns yellow. Addition of alkali or concentrated ammonia changes the color to orange. The reaction is caused by the presence of aromatic groups (phenylalanine, tyrosin, tryptophan).

The biuret reaction. A color varying from purple to violet develops when a protein solution in strongly alkaline medium is treated with dilute copper sulfate. The reaction is characteristic of substances containing two —NH—CO— groups joined directly or separated by a carbon or nitrogen atom. It is therefore induced by the peptide bond but can also be given by non-protein substances possessing such structures (e.g. biuret).

—NH—CO—RCH—NH—CO—	NH_2—CO—NH—CO—NH_2
peptide bond	biuret

The reaction may be used in quantitative analysis.

Ninhydrin (triketohydrindene) gives a blue color when heated with proteins, peptones or free amino acids. This reagent is widely used both for quantitative and qualitative determination of amino acids. *Folin's reaction* occurs when proteins are heated with a reagent containing β-naphthoquinone sulfonate. A red color is formed. A variation of this method serves in the quantitative spectrophotometric determination of proteins and amino acids.

2-5-4. *Hydrolysis.* The peptide bonds in proteins are hydrolyzed by strong acids, alkalis and certain enzymes. When heated for several hours in 6N HCl, proteins undergo complete hydrolysis into their component amino acids. Complete hydrolysis is also achieved by boiling with 5N NaOH. Hydrolytic cleavage of the peptide bond is catalyzed by proteolytic enzymes. Each proteolytic enzyme has its own specificity, i.e. it attacks certain types of peptide bonds preferentially (Section 2-8). Some degree of specificity is also observed in acid hydrolysis. Under certain conditions, treatment with dilute acid causes preferential cleavage of peptide bonds formed by aspartic acid, serine and threonine. The amide groups of glutamine and asparagine are also rapidly hydrolyzed.

Partial hydrolysis of proteins yields short peptide chains, containing a few amino acid residues. Small *peptides* may also be obtained by synthesis, starting with amino acids. Synthetic peptides are valuable in basic protein research. "*Peptone*" is a practical term used in older texts to indicate mixtures of small peptides obtained by partial enzymatic digestion of proteins.

A large number of simple peptides are found in nature. *Glutathione* is a tripeptide consisting of glutamic acid, cysteine and glycine.

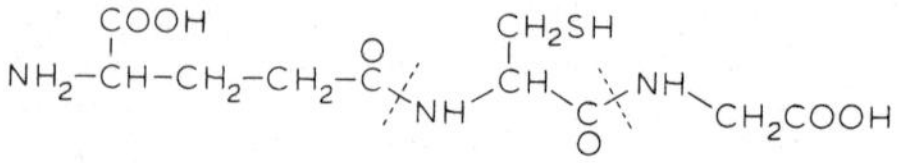

$$NH_2-CH(COOH)-CH_2-CH_2-C(=O)-NH-CH(CH_2SH)-C(=O)-NH-CH_2COOH$$

glutathione

Unlike in most peptides, in glutathione it is the γ-carboxyl group of glutamic acid that participates in the peptide bond. *Carnosine* and *anserine*, found in muscle, are both dipeptides of histidine and alanine. Small peptide groups participate in the molecule of penicillin and some other antibiotics. Some hormones are peptides.

The ability of strong acids to hydrolyze proteins into amino acids and small, water soluble peptides is used in the food industry for preparing the so-called *protein hydrolyzates* from plant materials. In this process proteins from various plant sources are hydrolyzed by strong hydrochloric acid, usually in glass-lined apparatus, and the product is then neutralized with alkali. The resulting mixture of peptides and amino acids is then concentrated and sold in this form as protein hydrolyzates. These are widely used for soup mixtures and similar food preparations.

2-5-5. *Oxidation–reduction.* Most susceptible to oxidation in a protein molecule are the SH groups of cysteine. Even in the presence of mild oxidizing agents the following reaction between two sulphydryl groups occurs:

$$R{-}SH + HS{-}R' \underset{[2H]}{\overset{[O]}{\rightleftharpoons}} R{-}S{-}S{-}R'$$

The reaction is generally reversible and the disulfide bond is broken by reducing agents. The formation of disulfide bonds from two sulfhydryl groups of the same molecule or from two different molecules (cross linkage) has an effect similar to that of vulcanization in rubber. It increases the rigidity of the protein. This effect explains the use of bromates and other oxidizing agents in bread baking, in order to increase the mechanical strength of the wheat protein, gluten (see Section 5-4).

Reducing agents such as cysteine, mercaptoethanol, glutathione are added to protein solutions when disulfide cross-linkage is to be cleaved or prevented.

2-5-6. *Colloidal and surface properties.* With molecular weights between 10^4 and 10^6, proteins fall within the range of colloidal particles. In solution, protein molecules may undergo association and thus form aggregates or micelles. Globular proteins adsorb water and swell considerably. The principal site of water adsorption is the peptide bond.

The viscosity of protein sols obeys Einstein's equation for the viscosity of a suspension where the dispersed phase is in the form of spherical particles.

$$\eta_s = \eta_m(1 + 2.5\phi)$$

η_s is the viscosity of the suspension, η_m that of the solvent, ϕ the volume fraction of the dispersed component. This equation may be used either to estimate deviations from sphericity or to determine the extent of solvation.

Due to the amphoteric character of proteins, the electrostatic charge of the colloidal particle in protein sols depends on the pH. Thus, it is possible to precipitate proteins from their sols by means of a colloid or large ion of the opposite charge. Use is made of this effect for the recovery of proteins from industrial waste liquors. Proteins adsorb ionizable dyes from their solutions. Adsorption on wool is sometimes used as a quick test for some dyestuffs in foods. Adsorption of a particular dye, *amido-black*, is a convenient quantitative test for protein in milk.

Solutions of proteins and particularly albumins tend to froth copiously. Foaming of egg white is the most important property of this material as an ingredient in confectionery and cake manufacture.

As long chains capable of hydrogen bonding and hydration, proteins can form quite stable *gels*, provided that the chains are expanded and interchain linking is possible. *Gelatin*, a protein obtained when collagen is subjected to the action of hot water, is a jellying agent, used extensively in food industries (see Section 5-1).

Protein molecules have at the same time hydrophobic side groups (such as the unsubstituted hydrocarbon radicals of alanine, leucine, etc.) as well as hydrophilic centers. Therefore, proteins can be expected to act as stabilizers of fat-water emulsions. A number of such emulsions are stabilized in nature by proteins. The fat globules in egg yolk and milk are surrounded by a "membrane" consisting mainly of lipoproteins.

2-5-7. *Sensory characteristics.* Pure proteins are usually tasteless, odorless and colorless. (An interesting exception is *monellin*, a recently characterized protein with an intensely sweet taste and a molecular weight of 10,000.) Any taste, color or odor associated with protein preparations come either from impurities or from prosthetic groups. When foods rich in protein undergo decomposition, putrid odors are formed. These are due to low-molecular weight decomposition products containing nitrogen, sulfur or both. Sulphydryl groups are easily detached from proteins, with the formation of H_2S (rotten egg). The characteristic odor of dough is partially due to the formation of H_2S from wheat gluten during kneading.

2-5-8. *Purification of proteins.* In nature, proteins occur in heterogenous mixture with many other substances. In research, it is often necessary to separate a protein from such mixtures. Various methods are available. *Dialysis* and *ultrafiltration* are useful for removing solutes of low molecular weight. Precipitation of proteins from aqueous solutions by acetone or by salting-out is a common practice. The separation of one protein from another is done by means of techniques which take advantage of differences in molecular size, shape and charge. These include ion exchange chromatography, gel filtration, preparative electrophoresis, isoelectric focusing. As separation techniques are perfected many proteins which were previously thought to be homogenous are found to consist of a number of different species.

2-5-9. *Crystallization.* Since the crystallization of hemoglobin in 1840, many soluble proteins have been prepared in the crystalline form. These include seed globulins, ovalbumin, several enzymes, protein hormones (insulin) and viruses. Scleroproteins show X-ray diffraction patterns indicating a considerable frequency of centers of crystallization. However, scleroproteins cannot be crystallized completely, because of their insolubility.

2-6. STRUCTURE OF PROTEINS

Molecular weight, amino acid composition and the linear sequence of amino acid residues along the peptide chain constitute what is known as the "*primary structure*" of a protein molecule. The spatial arrangement of the various atoms and groups in a molecule is referred to as "*conformation*".

2-6-1. *Molecular weights.* Here are a few examples of the approximate molecular weights of proteins:

insulin	5,700	β-lactoglobulin	35,400
ribonuclease	12,700	egg albumin	42,000
lactalbumin (cow)	17,500	hemoglobin	68,000
trypsin	24,000	urease	480,000
gliadin	27,000	myosin	620,000
zein	35,000		

Methods used for the determination of the molecular weights of proteins include the following:

(a) osmotic pressure
(b) light scattering
(c) viscosity, diffusion
(d) gel filtration
(e) sedimentation in an ultracentrifuge
(f) chemical composition.

Each of the above methods has its inherent advantages and disadvantages. The sedimentation, diffusion and viscosity methods, for example, all depend on the determination of the resistance of the molecules to their motion through the surrounding liquid medium. This resistance, however, depends not only on the molecular weight, but also on the shape of the molecules and the amount of water which moves with the molecule, i.e. hydration. Unless the shape and hydration are independently determined, the determination of the true molecular weight is impossible by any one of these methods. The light scattering methods are similarly affected by shape factors.

The osmotic pressure is theoretically independent of shape and hydration of the molecules, but the determination of molecular weight by this method is complicated by electrochemical factors (Gibbs-Donnan effect) and by deviation from ideal solution behavior.

The "minimum" molecular weight of a protein can be determined from its chemical composition on the basis of the fact that each molecule of protein must contain at least one residue of any of its amino acids or one molecule of its prosthetic group.

2-6-2. *Amino acid composition.* The determination of amino acid composition of a purified protein sample requires the following operations:

(a) Complete hydrolysis of the protein to its individual amino acids (see paragraph 2-5-4).

(b) Separation of the amino acids, usually by some chromatographic technique.

(c) Quantitative determination of each amino acid separately, usually by means of a color reaction (ninhydrin is frequently used).

The method of hydrolysis most frequently employed is that of acid hydrolysis, whereby the protein is heated with concentrated

(6N) hydrochloric acid. Tryptophan is destroyed extensively by this treatment and must be determined separately. Alkaline hydrolysis may be used for the determination of tryptophan but affects many other amino acids. Complete hydrolysis by enzymes is usually difficult and too slow.

Ion exchange chromatography is the prefered procedure for the separation of amino acids. Automatic *amino acid analyzers*, where the amino acids of the hydrolysate are eluted, measured and recorded continuously, permit the complete analysis of a protein sample in a matter of a few hours. Chemical conversion of amino acids to more volatile derivatives and subsequent separation and determination of these compounds by gas chromatography may well become a convenient and rapid method of amino acid analysis.

Amino acid composition is of particular interest to the nutritionist, since the nutritional value of a protein depends primarily on its amino acid pattern or *profile.*

2-6-3. *Sequential analysis. Dinitro-fluoro benzene* combines with the amino group of primary amines to give a *dinitro-phenyl* (DNP) compound:

$$O_2N\text{-}C_6H_3(NO_2)\text{-}F \; + \; H_2N\text{-}R \longrightarrow HF \; + \; O_2N\text{-}C_6H_3(NO_2)\text{-}NH\text{-}R$$

If a protein is treated with dinitro-fluoro-benzene, all the free amine groups (i.e. the α-amino group of the N-terminal residues as well as the ϵ-amino of lysine, etc.) are converted to DNP derivatives. If the protein is now hydrolyzed completely, these amino groups will retain their DNP "label" since the bond with DNP is resistant to hydrolysis. In this way the N-terminal amino acids of the protein can be identified. With a similar "labeling" technique specific to free carboxyl, the C-terminal amino acids can also be determined. It is easy to see how such reactions can be used to determine the sequence of amino acids in a dipeptide or tripeptide. Sequential analysis of a protein involves partial enzymatic hydrolysis of the molecule to small peptides, labeling, separation and identification of the latter and reconstruction of the protein model on the basis of the information gained. Different proteolytic enzymes must be used in order to obtain overlapping sequences. The first protein to be studied in this manner was *insulin.* In his brilliant work on the structure of insulin, Sanger found this protein to be composed of two peptide

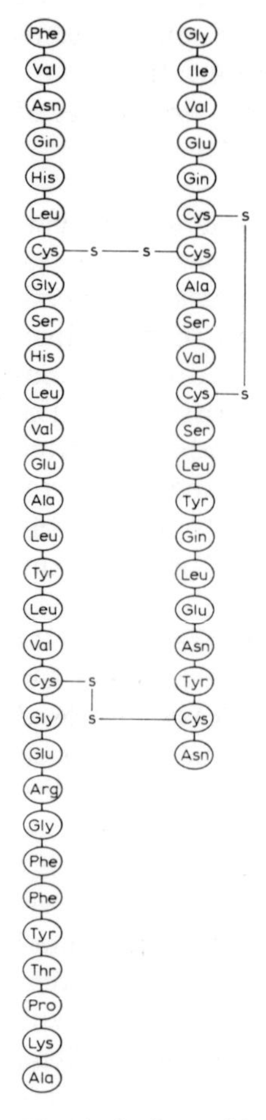

Fig. 1. Amino acid sequence of beef insulin.

chains joined by disulfide bonds. He then proceeded to isolate and identify many fragments and finally established the unequivocal sequence of the amino acids (Fig. 1).

The primary structure of the enzyme ribonuclease (124 amino acids), of the tobacco-mosaic virus (158 amino acids) and of hemoglobin (574 amino acids) and many other proteins has been elucidated since then. In the light of the information in hand, the following observations can be made:

(a) Each protein has a constant composition of amino acids arrayed following a fixed sequence, particular to that protein.

(b) The particular sequence is often the clue to the properties of the protein and especially to its biological activity.

2-6-4. *Conformation.* A large molecule such as that of proteins could assume a multitude of different shapes, all satisfying the conditions set by the covalent structure. However, in most native proteins, the various atoms and groups about the polypeptide chain occupy certain prefered positions in relation to each other, thus creating a particular arrangement in space, or conformation. The term *conformation* includes both the relative position of neighboring groups (*secondary structure*) and the overall folding of the chain over itself (*tertiary structure*). A fourth level of organization, the *quaternary structure*, involves the non-covalent association of several polypeptide chains to form the protein particle.

Theoretical considerations as well as X-ray diffraction data indicate that the atoms of a peptide bond lie on the same plane. On the basis of the peptide bond structure, Pauling and Gorey proposed a number of stable secondary structures. They assumed that each C=O group forms a hydrogen bond with an NH-group. One of the ways to bring those two groups close together was to assume that the chain is coiled in helical fashion, a C=O group on one turn of the helix facing a NH-group on the next turn. Calculations led to the conclusion that a prefered configuration would be a right-handed *α-helix* with a pitch of approximately 5.4 Å, and 18 residues in five turns. This view received considerable experimental support, the α-helix being indeed the most commonly encountered helical conformation (Fig. 2).

Another stable conformation, which also satisfied the conditions of hydrogen bonding between each C=O and NH is that known as the *pleated sheet* or *β-structure.* Here, the hydrogen bonds are formed between two different chains running side by side. This conformation is common to a number of scleroproteins (Fig. 4).

Any conformation is stabilized by secondary valence bonds and occasionally by covalent bonds (disulfide linkage, ester linkage between serine-OH and carboxyl etc.). *Hydrogen bonds* are probably the most important stabilizing factor. The so-called hydrophobic bond between non-polar side chains contributes also to the delicate balance of forces which stabilizes the structure. This last bond can be

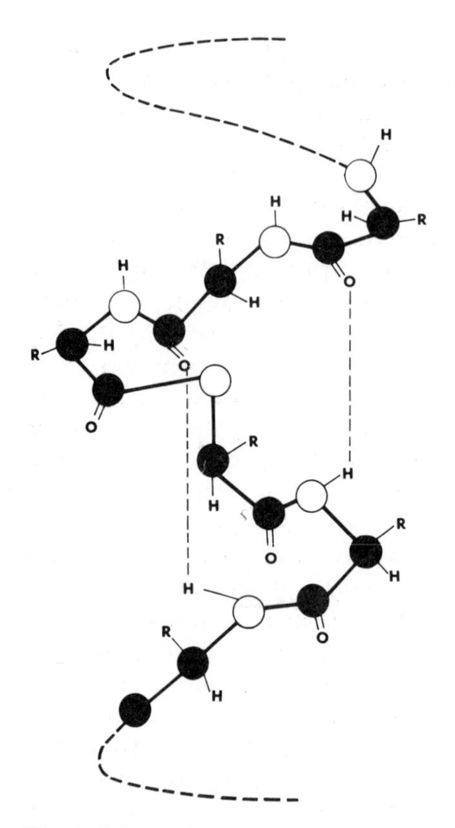

Fig. 2. Schematic representation of α-helix conformation.

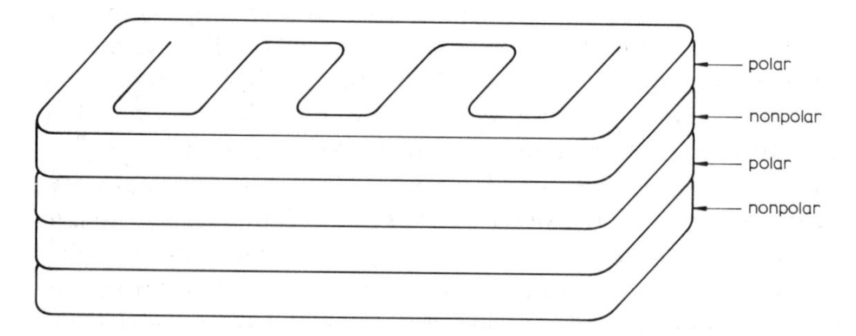

Fig. 3. Schematic representation of globular proteins.

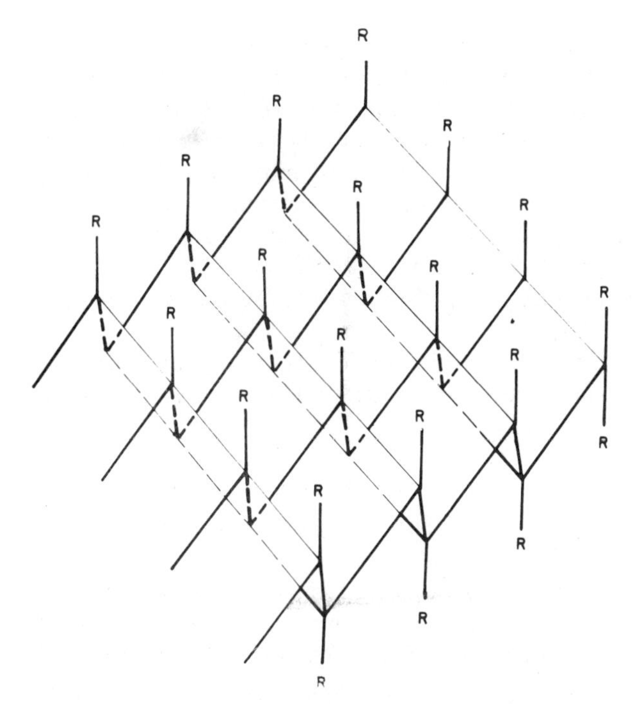

Fig. 4. Schematic representation of pleated-sheet structure.

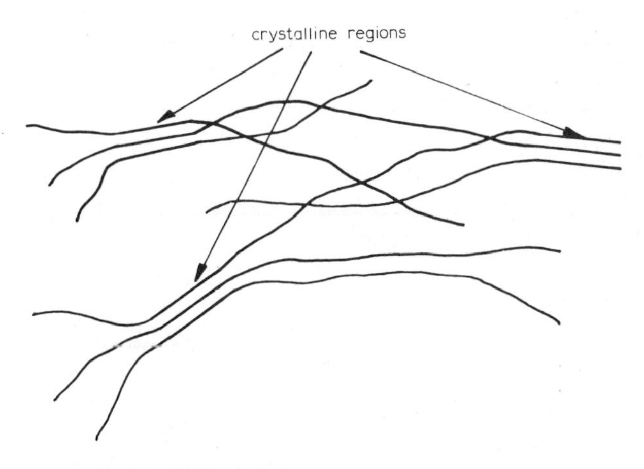

Fig. 5. Schematic representation of crystalline regions in protein.

regarded as the product of repulsion of the non-polar hydrocarbon side chains by polar, hydrophyllic groups or by the polar solvent itself. The net effect of all these forces is the creation of a highly ordered structure. In fibrous proteins, a number of helical chains are twisted so as to form fibers in the shape of multi-strand cables. In globular proteins the helix is folded in a particular way to form compact structures (Fig. 3). Most of the polar side groups are situated on the surface while the non-polar groups are hidden inside. Proline introduces an element of inflexibility and the helix is bent at the site of proline residues.

The formation of crystalline regions (Fig. 5) greatly increases the stability of tertiary and quaternary structures, especially in fibrous proteins.

2-7. DENATURATION

Heat, strong acids and bases, solvents such as ethyl alcohol, concentrated solutions of some salts, urea, a number of phenolic substances, usually bring about profound changes in the physico-chemical properties of soluble proteins. These changes include loss of solubility (coagulation), formation of irreversible gels, exposure of reactive groups such as sulfhydryl, higher susceptibility to enzymatic hydrolysis (and therefore better digestibility), alteration of X-ray diffraction patterns and optical rotation. The changes are often but not always, practically irreversible. It has been customary to group all these changes under one term: *denaturation.*

At the molecular level, denaturation is defined as any modification in the conformation of a protein. Thus, denaturation involves more or less extensive destruction of the secondary valence bonds responsible for the particular conformation of the native protein, but no rupture of covalent bonds, no proteolysis. Very often the net result of denaturation of a globular protein is the unfolding of the polypeptide chain and its transformation into a *randomly coiled* polymer (Fig. 6). This transition explains the formation of gels, coagulation, the increase in reactivity and susceptibility to enzymatic action. Colvin describes denaturation as an *order-disorder transition.* (The term *practically irreversible* is equivalent to saying that the *renaturation* process is much slower than the denaturation process.) Karlsow compares denaturation of proteins to the melting of a crystalline substance. Just as the orderly crystalline lattice is annihilated by melting, so is the high level of order in the

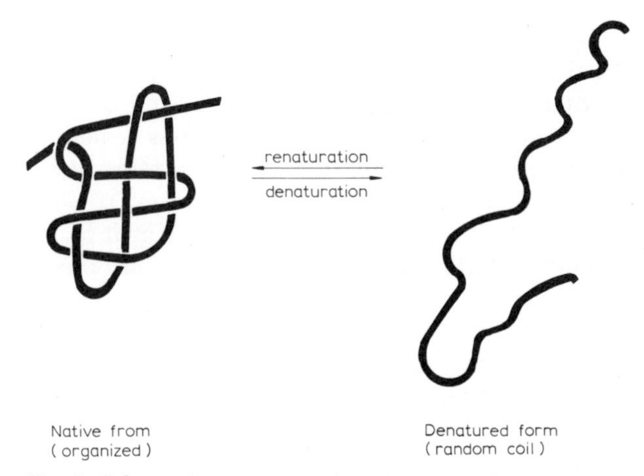

Fig. 6. Schematic representation of protein denaturation.

conformation of a native protein destroyed in the course of denaturation.

The usefulness of the term denaturation is debatable. This word is used to describe a multitude of phenomena, which may be fundamentally different one from another. Interpretation of the term is difficult in the case of proteins which do not have an orderly structure in the native state. Some proteins, such as caseins, are "born denatured". Nevertheless, the term "denaturation" conveys the idea of distinct changes in the properties of a protein and its use is legitimate as long as one also explains which particular changes are referred to.

Protein denaturation is a process of primordial importance in food technology. Thermal inactivation of enzymes (see Section 3-8) is largely due to denaturation. Protein denaturation may contribute to both the texture and flavor of many foods. Thus, exposure of —SH groups contributes largely to the specific flavor of cooked protein rich foods such as eggs and milk. Specific examples of denaturation in food proteins will be discussed in Chapter 5.

2-8. PROTEOLYTIC ENZYMES

As will be seen in the next chapter, *enzymes* are biological catalysts, themselves proteins with or without an additional non-proteinic moiety called *coenzyme* or *prosthetic group.*

A large number of enzymes, derived from various sources, specifically engaged in the breakdown of proteins are known today. These are named *proteases*, *peptidases* or proteolytic enzymes.

Proteolytic enzymes are classified into two major groups:

(a) *Endo-peptidases*, which are capable of attacking peptide bonds inside the chain.

(b) *Exo-peptidases* which hydrolyse peptide bonds adjacent to the terminal amino (*amino-peptidases*) or carboxyl (*carboxy-peptidases*) groups.

All of these enzymes are of relatively high specificity (see Chapter 3). Whereas the endo-peptidases can cleave proteins mainly into smaller peptide fragments, the products of exopeptidase activity are single amino acids.

A number of well known proteases are found in the gastro-intestinal tract of mammals:

Pepsin initiates the digestive breakdown in the stomach at the very low pH of about 1.0. This endo-enzyme, which has been crystallized from beef, salmon and tuna, causes the degradation of nearly all proteins; however, its specificity lies in its favoring the hydrolysis of the peptide bond at the amino side of an aromatic amino acid, such as tyrosine and phenylalanine.

Trypsin and *chymotrypsin* are two important proteases present in the mixture of enzymes found in the pancreas. Trypsin has also been crystallized. Its isoelectric point is near pH 8. Trypsin is an endopeptidase and it is specific towards peptide linkages formed by carboxyl terminals of amino acids of strongly basic character: arginine and lysine.

Chymotrypsin actually represents a group of several closely related endopeptidases (usually marked by the Greek letters α, β, γ, δ and π). The specificity of this group of proteolytic enzymes is similar to that of pepsin but their optimum pH is near 8.

All three proteases mentioned above, pepsin, trypsin and chymotrypsin, are respectively derived from their inactive precursors: *pepsinogen*, *trypsinogen* and *chymotrypsinogen*. The most interesting thing about this is that the conversion of these precursors into the active enzymes is catalyzed by the respective enzymes themselves; these reactions, therefore, are autocatalytic.

To complete the digestion of proteins in the gastrointestinal tracts of mammals, several exopeptidases are known which cleave the smaller peptide chains into amino acids.

Following death, animal tissues can undergo a process of "autolysis", which involves extensive degradation of their proteins into peptides and amino acids. The autolysis of such tissues is caused by enzymes named *cathepsins.* These are liberated and activated after death. The action of such enzymes in muscle is of considerable interest to the food industry in the matter of enzymatic tenderization of meat during aging (see Section 5-1).

To conclude the list of the numerous proteolytic enzymes derived from animal source, mention must be made of the enzyme *rennin*, probably known since ancient times, and found in the fourth stomach of the calf. Rennin causes the coagulation of milk casein by converting it into paracasein which is precipitated in the presence of suitable concentrations of calcium ions (see Section 5-2).

The next important source of proteolytic enzymes is the plant kingdom, where they are widely distributed in various plant tissues (*papain* from papaya, *bromelain* from pineapple, *ficin* from figs, etc.).

Finally, microorganisms constitute an important source of proteolytic enzymes, since bacteria, yeasts and fungi contain a large variety of proteinases and peptidases. Although the greater part of these have not been extensively purified, many of them are used in the food industry, especially in the ripening of special cheeses (Limburger, Camembert). In such processes the microbial culture functions as a source of extracellular proteinases which slowly hydrolyze the proteins during the aging of the cheeses to give them a smooth, buttery texture.

SELECTED BIBLIOGRAPHY TO CHAPTER 2

Bailey, J.L., Techniques of Protein Chemistry, 2nd edn., Elsevier, Amsterdam, 1967.

Dickerson, R.E. and Geis, I., The Structure and Action of Proteins, Harper and Row Publishers, New York, 1969.

Elmore, D.T., Peptides and Proteins, University Press, Cambridge, 1968.

Neurath, H., The Proteins, 3rd edn., Vols. I-IV, Academic Press Inc., New York, 1963-1966.

Sanger, F. and Tuppy, H., The Amino Acid Sequence in the Phenylalanyl Chain of Insulin, Biochem. J., **49**, 463-490 (1951).

Shultz, H.W. and Anglemier, A.F. (Eds.), Proteins and their Reactions, The Avi Publishing Co., Wesport, Conn., 1964.

Tanford, C., Protein Denaturation, in Advances in Protein Chemistry, Vols. 23 and 24, Academic Press, New York, 1968 and 1970.

CHAPTER 3

ENZYMES

In the yeast

KUHNE

3-1. DEFINITIONS

A *catalyst* is a substance capable of accelerating a chemical reaction, without itself undergoing any significant net change. *Enzymes* are proteins with the ability to catalyze specific biochemical reactions. Enzymes have been occasionally mentioned in the previous chapters, but since all known enzymes are proteins, it is here the appropriate place to go into further details regarding their constitution, properties and mechanism of action.

It was Berzelius who suggested in 1836 the term "catalysts" for substances that are able "to set into activity affinities which are dormant at a particular temperature by their presence alone". He ascribed these phenomena to some mysterious "catalytic power". This belief in such a mysterious power persisted for a long time, even after many of these enzyme preparations had been well established. As late as 1926, Willstaetter delivered a lecture before the German Chemical Society in which he summarized his researches on the isolation of pure enzymes, and in that lecture he related his work on *peroxidase*, which consisted of a protein and a non-protein moiety containing iron. The activity of this enzyme continued even when its concentration was diluted to such an extent that no more protein or iron could be detected in the substrate—this led Willstaetter to believe in a new natural force created in some way by the enzymes.

In the same year, 1926, Sumner succeeded in isolating and crystallizing from jackbeans the enzyme *urease*, which cleaves urea into CO_2 and NH_3, while Northrop and his collaborators crystallized the proteolytic enzymes *pepsin*, *trypsin* and *chymotrypsin*. Many scientists did not believe that these crystals were enzymes or that they even contained enzymes, and it was another 20 years before

Northrop, Stanley and Sumner received the Nobel prize for their brilliant achievements.

Many biological reactions can, of course, proceed in vitro as ordinary chemical reactions when relatively high temperatures or concentrations of acids or alkalis are applied. As already mentioned, proteins can be hydrolyzed into amino acids, for instance, by boiling for several hours in acids or alkalis, or again starch can be converted into glucose, etc. However, all such reactions proceed in vivo at body temperatures albeit with the active participation of enzymes.

The name "*enzyme*" was coined in 1878 by Kuhne from the Greek "*in yeast*". It was accepted at the time that enzymes could do their work only in the living cell. In 1897, the Buchners found for the first time that enzymic activity was not necessarily limited to the living cell: they triturated yeast cells with sand in a mortar and pressed out the "yeast juice", which, after thorough filtration in order to eliminate any cell particles, still preserved the ability to ferment sugar into alcohol and carbon dioxide. This was a great discovery at that time, for it separated for the first time biological phenomena from the concept of some mysterious "*vital power*".

Today, the activity of many hundreds of enzymes is acknowledged. More than a hundred enzymes have been crystallized. The amino-acid sequence and secondary structure of a few have been elucidated.

The substance, the transformation of which is catalyzed by an enzyme, is called the *substrate* of this enzyme.

Many enzymes are conjugated proteins. The non-protein moiety, termed "*prosthetic group*" is often only loosely attached to the protein and can be separated by dialysis. In this case the protein moiety is termed "*apoenzyme*" and the prosthetic group "*co-enzyme*" or "*cofactor*". The apoenzyme-coenzyme system is a "*holoenzyme*".

It is difficult and often meaningless to define the quantity of an enzyme present in a preparation in direct terms of concentration. Instead, the quantity is expressed as "*activity*" or the rate at which the specific reaction proceeds in the presence of the enzyme preparation, under standard conditions.

Activity units are arbitrary and their definition varies from one enzyme to another. Lately, there is a tendency to define one activity unit of any enzyme as *that quantity of enzyme which will transform its substrate at the rate of one micromole per minute, under standard conditions.* The "*specific activity*" of an enzyme preparation is the number of activity units per mg of protein.

3-2. CLASSIFICATION

There is no unanimity in classifying the many known enzymes and most of the textbooks use different systems: some arrange the enzymes according to the reactions they catalyze, others do it by the substrate on which the enzyme acts. Many enzymes are still designated by nonsystematic names coined to them by their discoverers or for historic reasons. In an attempt to introduce some standardization in enzyme nomenclature, the Commission of Enzymes of the International Union of Biochemists proposed a new systematic classification, which divides the enzymes into six main groups according to the type of reactions involved:

I. *Oxidoreductases* are enzymes engaged in the oxidation of biological systems. This group includes the enzymes commonly known as *dehydrogenases*, *reductases*, *oxidases*, *oxygenases* as well as *hydroxylases* and the enzyme *catalase.*

II. *Transferases* are enzymes which catalyze the transfer of various chemical groups (methyl, acetyl, aldehyde, ketone, amine, phosphate residues, etc.) from one substrate to another.

III. *Hydrolases* are responsible for the hydrolytic cleavage of bonds. Their trivial names are formed by adding the suffix *-ase* to the name of the substrate: e.g. *lipases*, *proteinases*, *pectinesterase*, *amylase*, *maltase*, etc.

IV. *Lyases* also catalyze the cleavage of bonds, but not by hydrolysis. They remove groups from their substrates, leaving double bonds. Examples are *carboxy-lyases* (decarboxylases), *aldolases*, *hydratases.*

V. *Isomerases* transform their substrates from one isomeric form to another. *Racemases* and *epimerases* are members of this group.

VI. *Ligases.* This name is now given to a group of enzymes which catalyze a certain type of synthesis.

Many of the enzymes mentioned above will be discussed in more detail in chapters dealing with their substrates.

3-3. COENZYMES

Earlier in this chapter, mention was made of Buchner's observation on fermentation of sugars by cell free yeast extract. In 1906,

Harden and Young fractionated yeast extract into a heat-stable, dializable component and a non-dializable, heat-labile protein fraction. None of the isolated fractions had the ability to ferment sugars, but this ability was restored when the two fractions were mixed again. The non-dializable component, which received the name of "*cozymase*" and later "*coenzyme I*" is a nucleotide (Section 6-4-4): *nicotinamide-adenine dinucleotide (NAD).*

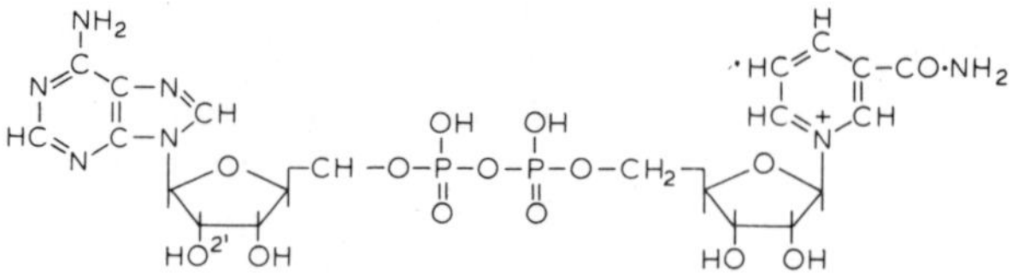

NAD (nicotinamide-adenine dinucleotide)

With an extra phosphate group on the hydroxyl marked 2′ the same combination constitutes another coenzyme: *nicotinamide-adenine-dinucleotide phosphate* (NADP). In older texts NAD and NADP are named *diphosphopyridine nucleotide* (DPN) and *triphosphopyridine nucleotide* (TPN), respectively.

NAD and NADP serve as cofactors to a large number of enzymes of the oxidoreductase group. These enzymes, known as *dehydrogenases,* oxidize their substrates by removing hydrogen from their molecules and transferring it to the coenzyme NAD (or NADP). Reduced NAD is re-oxidized by means of another enzyme which transfers the hydrogen to still another substrate (usually another coenzyme).

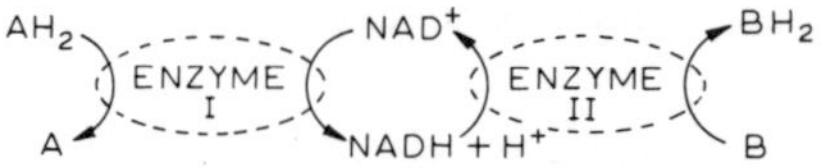

From this scheme it is evident that NADP serves as a substrate to its apoenzyme. In fact, the distinction between coenzyme and substrate is difficult, and coenzymes are often termed *cosubstrates.* The main difference between a true substrate (e.g. AH_2 in the scheme shown above) and a coenzyme is in their fate. In an enzymic reaction the true substrate is that compound which undergoes a net change, while the coenzyme follows a cyclic process of regeneration.

The property of NAD (or NADP) to serve as a reversible hydrogen acceptor-donor is located at the nicotinamide portion of the molecule:

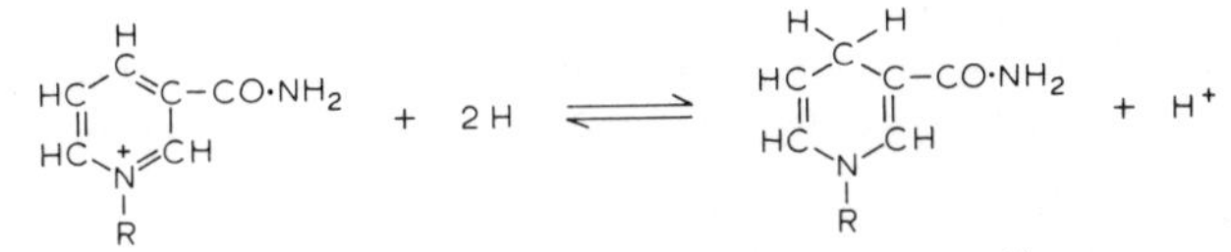

NAD^+ (oxidized) $\rightleftharpoons$ NADH (reduced)

$$\text{NAD}^+ + 2\,H \rightleftharpoons \text{NADH} + H^+$$

Another important group of nucleotide coenzymes are the flavin nucleotides. *Flavin mononucleotide* (FMN) is a component of the "old yellow enzyme" discovered by Warburg and Christian.

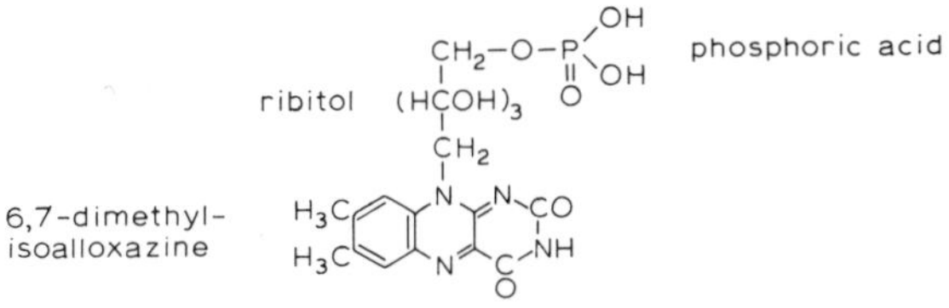

phosphoric acid

ribitol

6,7-dimethyl-isoalloxazine

formula of flavin-mononucleotide (FMN)

Flavines nucleotides are also "co-dehydrogenases". The enzyme-cofactor complexes are known as "flavoproteins". The most common co-factor in flavoproteins is, however, *flavine-adenine-dinucleotide* (FAD).

adenine moiety

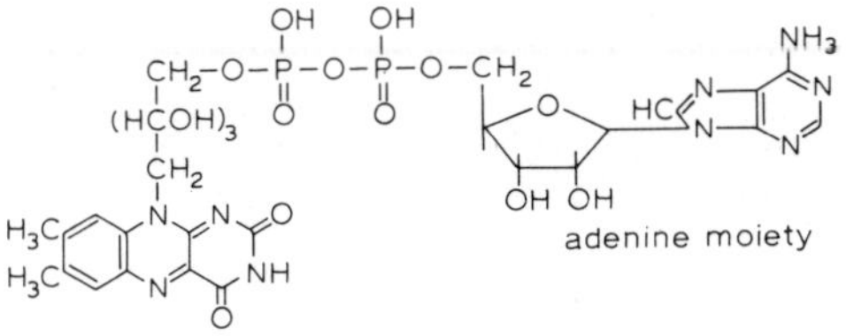

riboflavin moiety   formula of flavin-adenine dinucleotide (FAD)

The ability of flavin nucleotides to bind and carry hydrogen is associated with the isoalloxazine group of the molecule.

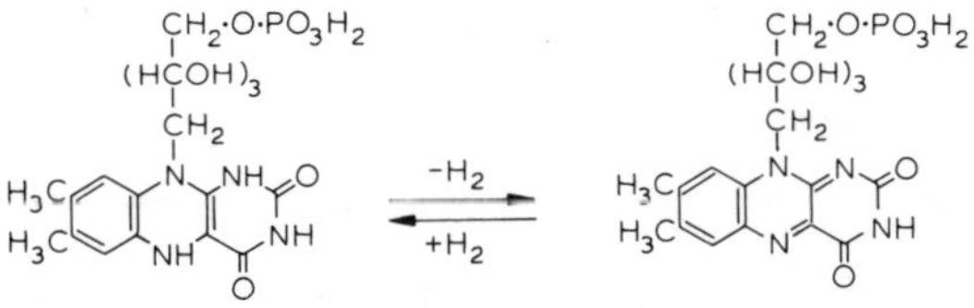

$FMNH_2$–(reduced enzyme) $\underset{+H_2}{\overset{-H_2}{\rightleftharpoons}}$ FMN–(oxidized enzyme)

Many co-factors are nucleotides. A number of nucleotides cannot be synthesized completely by the human body which lacks the enzymes necessary for the synthesis of certain parts of the molecule. Thus, in the case of NAD, the body cannot produce the nicotinamide group. "The missing part" must be supplied in the food. Therefore nicotinamide is an essential dietary factor which must be present in human food. Since NAD is regenerated almost completely in the enzymic reactions as shown above, only small amounts of nicotinamide are needed for the make-up of destroyed coenzyme and for growth.

Nicotinamide (niacin) is one of the many organic compounds which must be present in small amounts in food for the proper function of the body. Such compounds are termed *vitamins.* At least one large group of vitamins (the B vitamins) are known to serve in the same manner as nicotinamide, i.e. as sources of "missing parts" for the synthesis of important coenzymes.

Vitamin B_2 (riboflavin) is another example of a nutritional factor essential for the synthesis of co-enzymes, in this case flavine nucleotides.

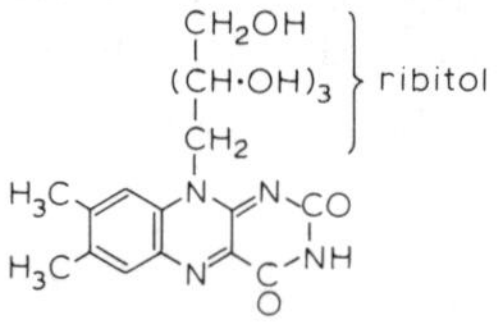

CH_2OH — $(CH{\cdot}OH)_3$ — CH_2 } ribitol

vitamin B_2 (riboflavin)
6,7-dimethyl-9(D-ribityl)isoalloxazine

3-4. MECHANISM OF ENZYMIC ACTION

The general mechanism of enzymic action is not different from that of all catalytic processes. At the first stage, the substrate has to combine with the catalyst, into a *substrate–enzyme complex.* This complex is unstable under the reaction conditions and is quickly decomposed. The enzyme is regenerated. The substrate, however, emerges from the reaction, either in the final, changed form, or in a highly activated state.

A discussion of the many possible mechanisms of activation are outside the scope of this book. The interested reader is referred to a text on the general chemistry of catalysis. We shall describe only one possiblity which seems to be extensively applicable to enzymic reactions.

Most biological reactions take place in dilute aqueous solutions. Assume two reactants, say a hydrogen donor A and a hydrogen acceptor B. The rate of reaction (transfer of hydrogen from A to B) is proportional to the frequency of collisions between A and B, which may be very low if the solution is dilute. Furthermore, only a small percentage of the collisions result in reaction. Now assume the presence of a catalyst C, capable of fixing both A and B on its surface. The concentration of the reactants on the surface of the catalyst would be much higher than in the solution. Furthermore, the catalyst may possess the ability to orient the adsorbed molecules in a specific way, so that the reactive groups of A and B become more available one to another. Such conditions would, of course, increase the rate of reaction between A and B. The concentration-orientation effect described above is only one of the possible mechanisms of catalysis.

The existence of an intermediary stage whereby the enzyme combines with its substrate to form an unstable complex has received ample experimental proof. The nature of the enzyme-substrate bond is not always known, but covalent binding has been demonstrated in a large number of cases.

The assumption that substrate binding is a prerequisite for catalysis explains two of the most characteristic properties of enzymes, namely their *specificity*, and their susceptibility to *inhibition* by certain substances.

3-5. SPECIFICITY OF ENZYMES

Enzymes display a degree of specificity seldom found in non-biological catalysis. It has been customary to distinguish between more specific enzymes and less specific ones but, as separation techniques became more and more accurate, enzymes which had been thought to possess a low degree of specificity were found to be actually mixtures of highly specific enzymes.

Reaction specificity implies that a given enzyme catalyzes only one of the various changes which can occur to a substrate. Thus a hydrolase is a specialized catalyst for hydrolytic reactions only. The fact that various enzymes of the same reaction specificity (e.g. dehydrogenases) require the same coenzyme (e.g. NAD) led to the conclusion that reaction specificity resides in the coenzyme while the apoenzyme carries substrate specificity. In the light of our present understanding of coenzymes as specialized transfer media (co-substrates), the function of reaction specificity also reverts to the primary catalyst, the apoenzyme.

Substrate specificity means that the enzyme is selective towards its substrate. There are several types of substrate specificity:

(a) *Steric specificity.* Many of the organic compounds taking part in the metabolic processes with which biochemistry is concerned are optically active substances and of the two stereo-isomeric configurations possible, only one is usually found in nature. It is interesting to observe that the known enzymes act only on one of these optical configurations and not on its enantiomorph.

Example: The hydrolytic enzyme, *arginase*, catalyzes the cleavage of the amino acid, L-arginine, into urea and L-ornithine, but will not do so with D-arginine:

NH NH_2 C $NH + H_2O$ $(CH_2)_3$ $CH{\cdot}(NH_2){\cdot}COOH$	arginase $\rightleftharpoons$	NH_2 NH_2 CO + NH_2 $(CH_2)_2$ $CH(NH_2){\cdot}COOH$
L-arginine		L-ornithine

Further example: The enzyme *lactic acid dehydrogenase*, isolated from muscle, and containing NAD as its prosthetic group, will catalyze the conversion of L(+)lactic acid into pyruvic acid, but it will not attack the D-form:

CH_2 $HOCH$ $COOH$	-2 H $\rightleftharpoons$ NAD	CH_3 $C{=}O$ $COOH$
L(+)·latic acid		pyruvic acid

Note that when the same enzyme acts on pyruvic acid, which is optically inactive, it will convert it into the L-form only; its enantiomorph will never appear with this enzyme. There is, however, another specific dehydrogenase, found in many microorganisms, such as *Bacillus delbruckii*, capable of forming only the D-lactic acid. This particular microorganism is used industrially in the production of lactic acid.

A third example of steric specificity is concerned with the cis-trans isomerism. *Succinic acid dehydrogenase* oxidizes succinic acid into *fumaric acid*, which has a trans configuration and never into maleic acid, which is the cis-form of the same molecule:

$$\begin{array}{l} CH_2COOH \\ | \\ CH_2COOH \end{array} \underset{}{\overset{\text{s.a. dehydrogenase}}{\rightleftharpoons}} \begin{array}{r} CH{\cdot}COOH \\ \| \\ HOOC{\cdot}CH \end{array} \qquad \begin{array}{l} CH{\cdot}COOH \\ \| \\ CH{\cdot}COOH \end{array}$$

succinic acid — fumaric acid (*trans*) — maleic acid (*cis*)

(b) *Other types of specificity.* The other types known depend primarily on the degree of specificity. This could be best explained with examples of hydrolases. Their general action can be represented by the following schematic formula:

$$A{-}B + HOH \xrightleftharpoons{\text{hydrolase}} AOH + BH$$

where A—B represents a molecule of the substrate built of two parts A and B and the bond between them. Accordingly, there are three types of specificity:

(I) When only the type of the linkage is important to the enzyme no matter what A and B are—*low specificity*;

(II) When the linkage and only one of the components of the molecule are important to the enzyme—*group specificity*; and

(III) When both moieties as well as the bond linking them are required by the enzyme to be able to act—*absolute specificity.*

(I) As an example of the low specificity type enzymes, one can take the *lipases*: practically all lipases can cleave the linkage between the acid and the alcohol in the lipid, as long as it is an ester linkage; they will equally easily cleave ethylbutyrate as well as a triglyceride.

(II) Some glycosidases are limited in their action to only a glycosidic bond, and some peptidases to only the peptide bond. However, in Chapter 2, when speaking of proteolytic enzymes, mention was made of the specific requirements of carboxypeptidase, the structural requisite of which is the presence of a free α-carboxyl group adjacent to the peptide linkage to be hydrolyzed:

$$\cdots CH(R){-}C(=O){\nearrow}NH{-}CH(R){-}COOH$$

This enzyme will not hydrolyze the peptide bond of a peptide of this type:

$$\cdots CH(R){-}NH{\nearrow}C(=O){-}CH(R){-}NH_2$$

for which a different peptidase, an *aminopeptidase*, is required. The enzymes carboxypeptidase and aminopeptidase are both examples of the second type of specificity-group specificity. This is the most common type among the enzymes.

(III) *Absolute specificity*. This is the case of an enzyme which necessitates that all three parts of the substrate upon which it acts should fit the requirements, namely, each of the moieties A and B as well as the bond between them. A suitable example for this type of absolute specificity is the true maltase, found in germinating barley, malt-maltase. This enzyme will attack maltose only and will not act upon any other glucoside. The name maltase is, therefore, the appropriate designation for it.

Similarly, the enzymes arginase or succinic acid hydrogenase, mentioned above in connection with steric specificity, are enzymes of absolute specificity. Arginase is specific for arginine and will not cleave the urea from such compounds as δ-N-methyl arginine or α-N-methyl arginine in which one of the amino groups is substituted by a methyl. Similarly, succinic acid hydrogenase is specific for succinic acid and not for malic acid.

The specificity of enzymes is explained by the strict requirements of amino-acid sequence and structure around the active sites (see Section 3-6), for the formation of an activated complex with the substrate. Obviously only proteins could offer such a degree of variation and specialization.

3-6. SITE OF ENZYME ACTIVITY

When an enzyme reacts with its substrate, only certain regions of the protein molecule, known as "*active sites*", participate in the process. Active sites consist of special groups of amino acid residues, brought close together by the sequence and the particular folding of the enzyme protein. One can expect to find within the active sites, one or more amino acid residues with reactive side-groups (SH, OH, NH_2, etc.). As stated before, the complete or partial amino acid sequence of several enzymes is known. Using various labeling techniques, it has been also possible to map the portions of the polypeptide chain which serve as active sites. The portion of the protein molecule outside the active regions does not participate directly in substrate binding and catalysis but it provides the backbone and support for the active sites. It determines the orientation and spacing of the active sites and thus affects indirectly the activity and specificity of the enzyme. Similarly the spatial arrangement of the protein molecule determines the orientation of active sites and creates steric conditions which favour the accommodation of certain substrates but rejects others.

3-7. ENZYME INHIBITION

The rate of enzymic reactions may be accelerated by the presence of certain substances (*activators*), or slowed-down by the presence of others (*inhibitors*). The most common mechanism of inhibition is that of *competitive inhibition.* Here, the inhibitor (*I*) is a substance sufficiently similar in structure to the substrate (*S*), so that it can also combine with the enzyme (*E*) to form the complex *EI*. The presence of *I* diminishes the number of active sites available for binding the substrate *S*. The rate of formation of the normal complex *ES*, and hence the overall rate of the normal enzymic reaction decreases. Obviously, the more stable the abnormal complex *EI*, the more effective the inhibition. If *EI* is completely stable, the enzyme becomes permanently unavailable for its normal function; it is *poisoned*. The mechanism is often pictured by the famous key-and-lock analogy. To open a door, the "key" (substrate) must fit the lock (enzyme). A "wrong key" (inhibitor), may fit the lock if sufficiently similar to the "right key". However, it will not open the door, but will become stuck and will block the lock altogether.

For example, the enzyme *succinic acid dehydrogenase* catalyzes the oxidation of succinic acid into fumaric acid. This enzyme is very specific and will not dehydrogenate any other substance. However, if *malonic acid* is added to the substrate, it will compete with succinic acid, due to their close similarity in structure, by combining with the enzyme and making it unavailable for its duty, while malonic acid itself will not be dehydrogenated. Here, malonic acid is the "key" which does not fit, for some reason the "lock" (succinic acid dehydrogenase) and, at the same time, interferes with the opening of the door (the true reaction).

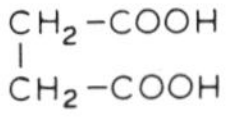

CH_2-COOH | CH_2-COOH	$CH_2(COOH)_2$
succinic acid	malonic acid

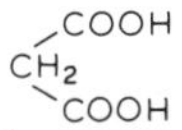

Inhibitors may block important metabolic pathways, in which case they are termed *antimetabolites.* Modern pharmacology uses anti-metabolite drugs extensively. For example, the well known *sulpha-drugs*, such as *p*-aminophenylsulphonamide, owe their therapeutic action to their structural resemblance with *p-aminobenzoic acid* (PABA), an essential factor for the growth of certain pathogenic bacteria.

NH_2 / COOH
p-aminobenzoic acid

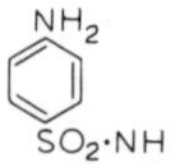

NH_2 / $SO_2{\cdot}NH$
p-aminophenylsulphonamide

In a similar way, it is possible to explain the action of some of the chemical or antibiotic preserving agents used in the conservation of foods.

A similar situation exists in the case of inhibition of the enzyme trypsin by some substances present in soybeans. Unheated soybeans are known to inhibit the proteolytic action of trypsin (see Chapter 2). This action is due to several low molecular weight proteins, known as *soybean trypsin inhibitors.* Trypsin inhibitors have been found in other food sources as well, such as several legumes and raw egg-white.

Trypsin is a proteolytic enzyme and the first stage of its action involves the formation of an enzyme-substrate complex. In the case of ordinary proteins this complex is unstable and the enzyme is constantly regenerated. However, the complex between trypsin and soybean trypsin inhibitors is a stable one. Once attached to its inhibitors, trypsin becomes unavailable for further proteolytic action.

3-8. INACTIVATION OF ENZYMES BY HEAT

As in the case with many proteins, enzymes can be easily denatured in several ways mentioned earlier, among them, heat. Most enzymes are, therefore, very thermolabile and it is usually sufficient to apply a temperature of 70 to 80° C for two to five minutes in order to destroy their activity.

This fact of complete inactivation of enzymes by heat is very widely used in the food industry. In most cases of food preservation, it is desired that there should be no continuation of any enzymic activity. A continuance of enzymic action may cause, for instance, a change in the color of the chlorophyll, or of the carotenoids, or cause browning of various foods; it may change the taste of carbohydrates, or cause rancidity in oils; it may cause changes in the flavor or in the nutritive value of proteins (or of vitamins) and, finally, the presence of pectolytic enzymes may cause complete changes in the texture of foods.

Heat treatment is, of course, a convenient method for the destruction of spoilage microorganisms in food (heat sterilization and pasteurization). Thus, proper thermal processing can achieve simultaneous microbiological preservation and enzymic stabilization of foods. Sometimes the residual activity of an enzyme (not necessarily objectionable) is used as a test for the adequacy of a heat treatment process. Thus, the absence of *phosphatase* activity in milk is a good indicator of whether the milk is adequately pasteurized, since this enzyme is completely inactivated by the dose of thermal treatment necessary to destroy pathogens such as B. tuberculosis.

Enzymes differ widely in their resistance to thermal inactivation. Plant peroxidases are particularly stable. Holding for a few minutes at temperatures as high as 120° C will not fully destroy heat-resistant peroxidase. The rate of thermal inactivation depends also on the pH, ionic strength and on the physical state of the enzyme in the food material: whether the enzyme is equally well distributed throughout the product or adsorbed on solid particles (as is the case with pectolytic enzymes and phenolases, which are adsorbed on the pulp of various fruit juices).

Situations are known in food technology where, although they had not been inactivated, some enzymes become "regenerated" and their activity renewed after some time. Such *enzyme regeneration* has been observed in the cases of peroxidases (milk, vegetables), catalase (vegetables), lipase (milk products) and pectolytic enzymes (citrus juices). Thermal inactivation of enzymes is apparently due to the disruption of active site structures as a result of denaturation. Regeneration would then be a result of reorganization, at least partial, of the protein molecule to restore the active sites. Reversal of denaturation is a slow process (see Section 2-7) but then sufficient time for measurable regeneration of some enzymes may be available during prolonged storage of processed foods. Thus the stability of food towards enzymic spoilage is a function of both the "depth" of thermal inactivation and storage time and conditions.

3-9. ENZYMES IN FOOD TECHNOLOGY

The significance of enzymic reactions in food technology constitutes the main part of the subject matter of the present book. Therefore, it would be impossible to treat the topic of enzymes in food technology within the framework of a single sub-title. Various enzymic processes including intentional use of enzyme preparations in the food industry will be discussed in connection with the substrates involved.

3-10. IMMOBILIZED ENZYMES

The industrial use of enzymes as catalysts has been limited, mainly for economic reasons. Enzyme preparations are expensive. Although the enzyme is regenerated in the process and could be used over and over again, its separation from the reaction mixture is not economically feasible.

Recently, it has been possible to attach enzymes to insoluble matrices. The enzyme thus immobilized has the advantages of a solid catalyst. It is easily separated from the reaction mixture. In some cases, especially when the supporting matrix is of ionic character, the properties of the fixed enzyme may be different from those of the soluble form.

Artificial immobilization of enzymes was developed and investigated in the early 1960s, but the industrial implementation of the idea is only starting. It is safe to assume that immobilized enzyme techniques will lead to successful industrial processes and will widen the scope of industrial catalysis by enzymes. One of the exciting developments in this field is the possibility of creating immobilized multienzyme systems by attaching more than one type of enzyme to the matrix. It should be noted that, in nature, endoenzymes, i.e. those operating inside the cells, are usually immobilized by fixation on membranes or on suspended solid particles.

SELECTED BIBLIOGRAPHY TO CHAPTER 3

Bernhard, S., The Structure and Function of Enzymes, W.A. Benjamin Inc., New York, 1968.

Boyer, P.D. (Ed.), The Enzymes, 3rd edn., Vols. I—VI, Academic Press, New York, 1970-1972.

Gutfreund, H., An Introduction to the Study of Enzymes, Wiley, New York, 1965.
Haldane, J.B.S., The Enzymes, M.I.T. Press, Cambridge, Mass., 1965.
Northrop, S.H., Kunitz, M. and Herriott, R.M., Crystalline Enzymes, 2nd edn., Columbia University Press, New York, 1948.
Reed, G., Enzymes in Food Processing, Academic Press, New York, 1966.
Whitaker, J.R., Principles of Enzymology for the Food Sciences, Marcel Dekker Inc., New York, 1972.
Wingard, Jr., L.B. (Ed.), Enzyme Engineering, Interscience Publishers, New York, 1972.

CHAPTER 4

ENZYME KINETICS

In this chapter we shall analyze the factors which affect the rate of enzymic reactions. However, before entering the subject, a short revision of the principles which govern chemical reaction kinetics is in order.

4-1. REACTION RATE

Consider the reaction:

$$A + B \rightarrow C$$

All other conditions (temperature, pressure) being constant, the velocity of the reaction (v), i.e. the rate of formation of C, is proportional to the concentration of A and B. Denoting these concentrations (in units, say, of molarity), as a and b, we have:

$$v = Kab \qquad (1)$$

K is the *rate constant* of this reaction. Its dimensions depend on the "order" of the reaction.

One of the most important factors which affect K is the temperature. The relationship between K and the temperature is given by Arrhenius' law:

$$K = Ae^{-E/RT} \qquad (2)$$

where:

A = constant
E = energy of activation
R = gas constant (1.99 cal/mol °C)
T = absolute temperature.

If K_1 and K_2 are the rate constants of the same reaction, but at the temperatures of T_1 and T_2, respectively, then:

$$\log \frac{K_2}{K_1} = \frac{E}{2.3R} \cdot \frac{T_2 - T_1}{T_1 T_2} = C\left(\frac{1}{T_1} - \frac{1}{T_2}\right) \qquad (3)$$

Furthermore, if the temperatures are not too far apart,

$$T_1 T_2 \cong T_1^2$$

It follows that

$$\log \frac{K_2}{K_1} = C'T \qquad (4)$$

In other words, within a narrow range of temperatures, a linear change in temperature causes a logarithmic change in reaction rate. The slope of the logarithmic line is determined by the pseudo-constant C' which contains E and the base temperature T_1. The well known concept of Q_{10} is based on this approximate relationship between temperature and reaction velocity. Q_{10} is the ratio between the rate constants at temperatures 10° C apart.

$$Q_{10} = \frac{K_{T+10°C}}{K_T} \qquad (5)$$

For most chemical reactions the value of Q_{10} is nearly 2.

4-2. ACTIVATION ENERGY

The mere presence of an enzyme and its substrate will not start a reaction. It is clear, therefore, that all enzymatic reactions can take place only if they are thermodynamically feasible. Whether the reaction will indeed take place is quite another question and depends

TABLE 1

Activation energy of some reactions with and without enzyme catalysis

Reaction	Enzyme	Activation energy—*Ea* (cal/mol)	
		No catalyst or with H^+ only	In the presence of the enzyme
Breakdown of H_2O_2	Catalase	18,000	5,000
Sucrose hydrolysis	Fructosidase	26,000	11,000
Casein hydrolysis	Trypsin	20,600	12,000
Lipid hydrolysis (ethyl butyrate)	Lipase	13,200	4,200

on certain conditions. The first and most important condition is that the molecules entering the reaction must be in an activated state.

It is the function of enzymes, as catalysts, to reduce the *activation energy* to a great extent. Table 4-1 shows the differences between the activation energy in various reactions without enzymes in comparison with those taking place in their presence.

In all these instances the activation energies required in the presence of enzymes are considerably smaller than in their absence or when the reaction is activated by the hydrogen ion alone, and because of the exponential character of the relationship between the rate of reaction and the activation energy, it is not difficult to calculate that a decrease from 20,000 to 10,000 cal/mol may correspond to an increase in the rate of reaction of about 500,000 times.

4-3. INFLUENCE OF TEMPERATURE

The second factor which may considerably influence the rate of a reaction, as shown by the Arrhenius' equation, is temperature. Although the general range of temperatures suitable for enzymic reactions is very narrow, slight changes have considerable influence. The *optimal temperature* for most enzymic reactions, with a few exceptions, lies between 30° C and 40° C. With the rise of temperature the reaction rate increases and for most enzymes, a rise of 10° C will double or even triple the rate of the reaction. However, the effect of temperature on enzyme activity is more complex than in the case of non-catalyzed reaction. A rise in temperature will cause an increase in reaction rate in accordance with Arrhenius' equation, but on the other hand, the same rise will accelerate inactivation of the enzyme by thermal denaturation. This dual action of heat may be expected to result in a temperature-activity relationship which can be expressed as a bell-shaped curve. This is confirmed by experiments. Under a given set of conditions each enzyme has an optimum temperature, at which its activity is at maximum (Fig. 7). For many enzymes, the region of extensive thermal inactivation lies very near to the optimum temperature.

At temperatures between 70° C and 90° C most enzymes are rapidly inactivated. On the other hand at low temperatures enzymic activity proceeds very slowly but is not entirely arrested. This should be taken into consideration in the frozen food industry: unless the

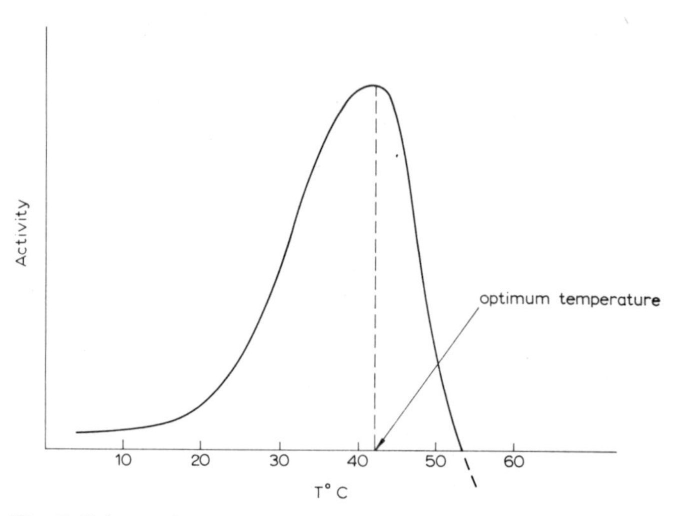

Fig. 7. Schematic temperature-activity curve.

objectionable enzymes are inactivated previously, many frozen foods will undergo considerable deterioration after prolonged storage, for even at such temperatures as low as —18° C (0° F) some enzymic reactions continue to take place.

The enzyme catalyzed reactions have activation energies in the order of 10 kcal/mol, while activation energies of thermal inactivation of enzymes are in the 50-150 kcal/mol range. This explains the assymetric form of the curve depicting reaction rate vs temperature. It also indicates that the optimum will depend on the incubation time. For longer incubation periods the optimum will occur at lower temperatures. Thus, the temperature optimum is not a true characteristic of an enzyme but a useful operational concept. It is meaningful only as long as the experimental conditions are well defined.

The effect of temperature on the rate of enzyme catalyzed reactions cannot be explained solely on the basis of Arrhenius plots for the reaction and for the thermal inactivation of the enzyme. The substrate itself may be affected by the temperature. Furthermore, if one of the reactants or one of the reaction products is a gaseous compound, temperature will affect its solubility in the reaction medium and therefore its rate of adsorption into or removal from the system.

4-4. INFLUENCE OF pH

The catalytic action of enzymes is usually attained within comparatively narrow limits of pH (Fig. 8). Every enzymic reaction has its optimal pH; however, the pH optimum depends on the substrates involved and on the reaction conditions (incubation time and temperature, substrate concentration, ionic concentration, etc.).

The effect of pH on the rate of enzyme catalyzed reactions is very complex. In the first place, active sites can be expected to contain ionizable side groups such as free —COOH, free $-NH_2$, —SH, —OH. The extent of ionization of these groups and therefore the structure and reactivity of the active site will obviously depend on pH. Similarly, the substrate itself usually contains ionizable groups and is therefore affected by the pH. At a second stage, the pH could also have a marked influence on the structure and reactivity of the enzyme-substrate complex, any inhibitors or activators present, etc.

Very high or very low pH may be expected to cause irreversible denaturation of enzymes. In fact, most enzymes lose their activity after exposure to pH values either much lower or much higher than the optimal range. The pH stability of enzymes varies with the temperature, ionic concentration and the purity of the enzyme preparation.

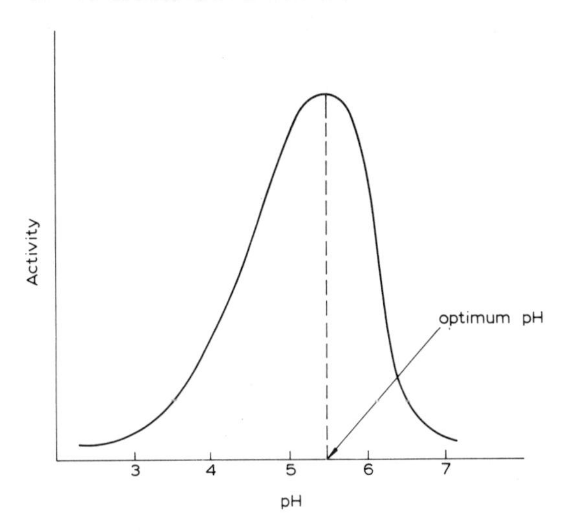

Fig. 8. Schematic pH-curve activity.

The influence of pH on enzyme activity is often exploited in food technology for retarding objectionable enzymic reactions or accelerating desirable ones by controlling the pH of the medium.

4-5. ENZYME CONCENTRATION

When all other factors comply with the optimal requirements, the rate of an enzymic reaction depends on the concentration of the enzyme. In fact, as long as the substrate is available in large excess, the reaction velocity is directly proportional to the enzyme concentration, as shown by the curve presented in Fig. 9, showing the inversion of sucrose by invertase.

4-6. SUBSTRATE CONCENTRATION

The rate of an enzymic reaction will, of course, normally rise with the concentration of the substrate, however, only up to a certain degree. From that point, the reaction velocity will change very little with the concentration of the substrate and ultimately the curve will

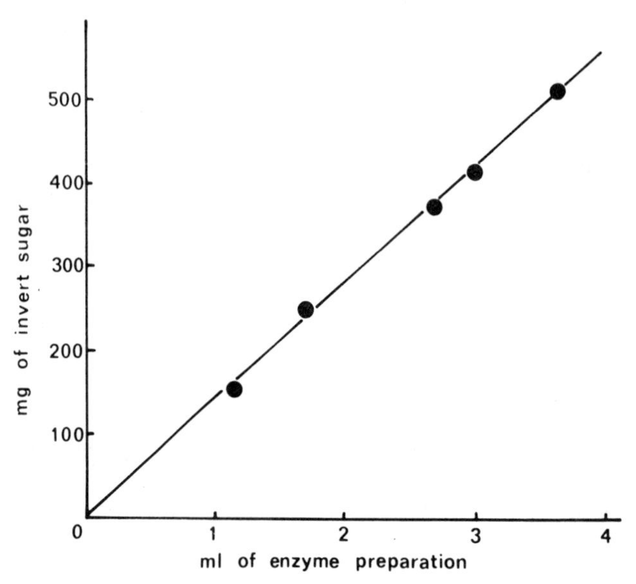

Fig. 9. Rate of reaction versus enzyme concentration.

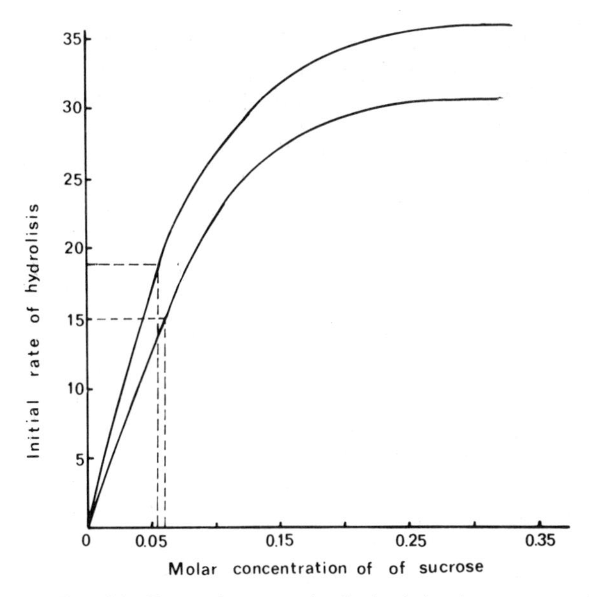

Fig. 10. Rate of sucrose hydrolysis by invertase at different concentrations of the substrate.

flatten out, i.e. the rate will become constant. With more enzyme in the substrate, a higher curve will be obtained but this will follow the same pattern. Figure 10 shows the results of an experiment carried out with invertase acting upon sucrose (with higher concentration of enzyme the hyperbolic curve was higher).

The experimental curves shown in Fig. 10 can be derived mathematically, in the light of the enzyme-substrate complex mechanism mentioned before. Consider the formation of the complex *ES* from the enzyme *E* and its substrate *S*, as the first step of the enzymic transformation of *S* to the end products P_1, P_2 etc.

$$E + S \underset{K_2}{\overset{K_1}{\rightleftharpoons}} ES \xrightarrow{K_3} P_1 + P_2 + \ldots$$

K_1, K_2, K_3 are rate constants of the individual reactions represented by the corresponding arrows. Let e, s, es be the concentration of E, S, ES at a given moment, and e_0 the initial concentration of the enzyme. Then:

$$e = e_0 - es \qquad (6)$$

By virtue of the Mass Law, the equilibrium constant of the complex formation reaction, K_m, is given by:

$$K_m = \frac{e \cdot s}{es} = \frac{(e_0 - es)s}{es} \qquad (7)$$

The overall velocity of the reaction, i.e. the rate of product formation, is:

$$v = K_3 \cdot es \qquad (8)$$

It is clear that if the concentration of the substrate is increased beyond a certain point, practically all the enzyme will be bound in the complex form and the concentration of free enzyme will be negligible. Under such conditions of saturation, $es = e_0$, which is the highest value es can attain. Since the velocity v is proportional to es, the condition of saturation also corresponds to the maximum reaction velocity. Equation (8) becomes:

$$v_{max} = K_3(es)_{max} = K_3 e_0 \qquad (9)$$

substituting equation (7) in equation (8), we obtain:

$$v = K_3 \frac{e_0 s}{K_m + s} \qquad (10)$$

which is equivalent, by virtue of equation (9) to:

$$v = v_{max} \frac{s}{K_m + s}$$

This is the Michaelis-Menten equation. K_m is a meaningful parameter of enzymic reactions and is known as the *Michaelis-Menten constant.*

Figure 11 shows theoretical curves based on the Michaelis equation and it is of interest to note that these correspond very closely to the curves obtained from actual experiments (see Fig. 10). This equation gives a rectangular hyperbolic curve with the following characteristics:

(a) The limit of the reaction velocity is an asymptotic value to which the velocity approaches with rise of the substrate concentration.

(b) The Michaelis constant, K_m, corresponds to that substrate concentration at which the velocity of the reaction reaches one-half of its maximum rate.

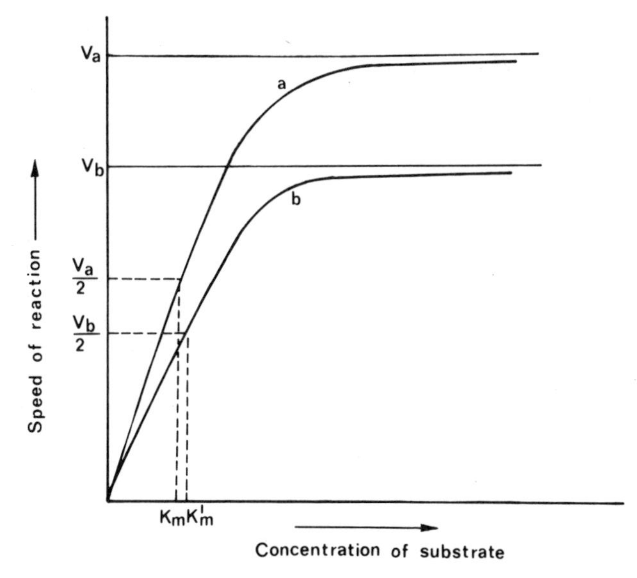

Fig. 11. Theoretical reaction velocity curves based on Michaelis equation.

(c) In the case of a large excess of substrate, the limit velocities are directly proportional to the concentration of the enzyme (as discussed earlier), and as can be seen from the differences in the heights of curves a and b.

(d) The constant K_m is a measure of the affinity between an enzyme and a substrate, and hence, of the stability of the enzyme-substrate complex. A large value of K_m represents low affinity and weak substrate-enzyme association. Thus, a substrate which has a low Michaelis constant with respect to a given enzyme, will be an efficient competitive inhibitor to another substrate with a higher K_m.

So far we have ignored the effect of the products P_1, P_2, etc. on the reaction rate. Obviously, as in any chemical reaction, accumulation of the products tends to slow down the reaction. Regulation of reaction rates by the concentration of reaction products is an important control mechanism in biology. This effect should be kept in mind in the design of technological processes as well. For instance, the hydrolysis of the bitter principle of grapefruits, naringin, by the enzyme naringinase, is depressed by the presence of sugars, which themselves are products of the hydrolysis.

4-7. INFLUENCE OF WATER ACTIVITY

In foods as well as in any biological system, water is one of the most important components. As a solvent, water serves to bring together the various interacting molecules. Furthermore, the reactivity of many substances depends on ionic dissociation and molecular configuration, hence on hydration. Water itself is often one of the reactants or one of the products of the reaction.

The influence of water on the rate of enzymic and non-enzymic processes in food has been recognized empirically for a long time. The possibility of retarding enzymic, microbiological or chemical deterioration simply by reducing the water content of foods was known to ancient civilizations. We shall limit our discussion here to the effect of water on the rate of enzyme catalyzed reactions. We shall start by reviewing some basic concepts regarding the state of water in foods and in biological systems.

The water content of foods varies from over 90% in some vegetables and fruits to a few percent in some grains and dehydrated foods. However, it is now recognized that the influence of water on the reactivity of the system is related not only to the actual water content but also to the state of the water molecules. The availability of water is a function both of *content* and *state* and is best expressed by means of a concept known as "*water activity*". Water activity α is defined as:

$$\alpha = \frac{p}{p_0} \tag{11}$$

where

p = water vapor pressure over the system
p_0 = vapor pressure of pure water at the same temperature.

If foods were mere mixtures of water with inert substances not interacting in any way with water molecules, the water activity would always be 1, regardless of the water content.

If foods were "ideal solutions" of solutes such as sugars, acids and salts in water, the vapor pressure would be governed by *Raoult's Law* and would be given by the equation

$$p = xp_0 = (1-c)p_0 \tag{12}$$

where x is the concentration of water and c is the total concentration of the solutes, both expressed in terms of mol fraction. This relationship depicts reasonably well the behavior of high moisture foods such as vegetables, fruits, juices, beverages, milk and fresh meat.

In food systems, part of the water is strongly adsorbed on the surface of polymeric substances (proteins, macromolecular carbohydrates). This is evident from the fact that the water vapor pressure over a food with low or intermediate moisture content is considerably lower than that predicted by Raoult's Law. This is known as "*bound water*".

An experimental plot of water activity against moisture content at constant temperature is termed a "*water sorption isotherm*". A sorption isotherm, typical of many foods, is illustrated in Fig. 12. It is believed that at very low levels of moisture content (region A of the isotherm) all the water is bound to the exposed polar sites of the macromolecular components (hydroxyls in carbohydrates, peptide bonds and polar side groups in proteins). A definite amount of water is needed to occupy all such sites and to form a "monolayer". This is usually known as the *BET monolayer*, because the theoretical calculation of its value was first proposed by Brunauer, Emmet and Teller in 1938. BET monolayer values for most foods are in the range of 3 to 10 grams of water per 100 grams of dry substance. Once the

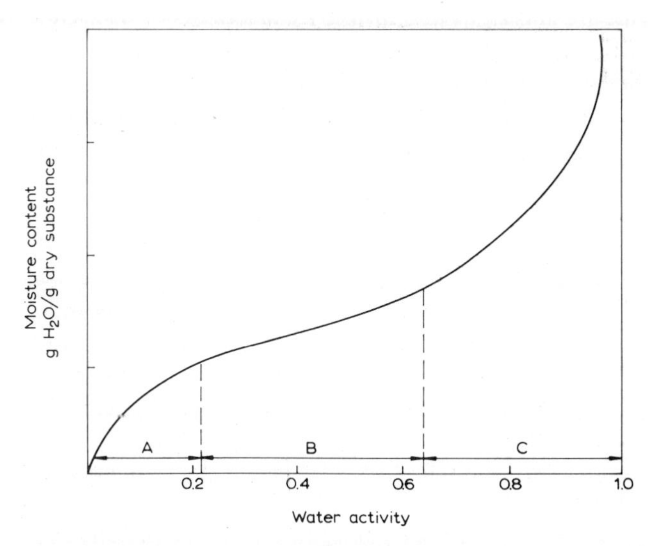

Fig. 12. Typical sorption isotherm.

monolayer is completed water activity increases sharply with increasing moisture content (region B). However, water activity is still lower than the value corresponding to concentration by virtue of Raoult's Law. One of the reasons for this effect is the fact that *free water* is partly entrapped in the porous structure of the food and therefore it is subject to the vapor pressure depressing action of capillaries (region B). At higher levels of moisture content (region C) the system actually behaves as an aqueous solution. In fact, the molar concentration of solutes is usually so low that water activity soon reaches values near unity.

The availability of water, measured as water activity, has a strong influence on the rate of enzyme catalyzed reactions, i.e. on enzyme activity. Acker has shown that the curve representing the activity of the fat splitting enzyme *lipase* as a function of water activity parallels the sorption isotherm of the experimental system. In general, enzymic activity is extremely low at moisture content levels below the monolayer value. Enzymic activity increases with increasing "free water" content. This relationship is found not only in hydrolytic reactions where water is one of the obvious reactants but also in non-hydrolytic reactions.

The role of water, other than active participation as a reactant, could be associated with the activation of the enzymes and substrates by hydration or with its action as a medium for transport. Observing that enzymic reactions involving non-polar substrates can take place even at moisture levels below monolayer values, provided that a suitable non-aqueous solvent is present in sufficient quantity, Acker concluded that the main role of water is that of a solvent, enabling the diffusion of the substrate to the enzyme.

4-8. KINETICS OF THERMAL INACTIVATION OF ENZYMES

As mentioned in Section 3-8, the thermal inactivation of enzymes is the basis of stabilization processes used frequently in food technology. Observations show that, as far as residual activity is concerned, this is a first order reaction. Therefore, when the logarithm of residual activity is plotted against time, a straight line is obtained for each temperature. The slope of the line is, of course, steeper at higher temperatures (Fig. 13).

The relationship between the rate of thermal inactivation and temperature obeys Arrhenius' law. The activation energies calculated

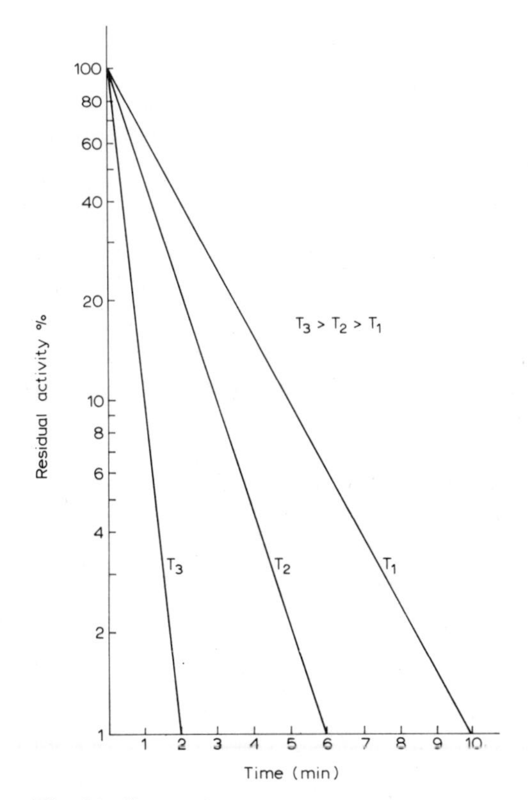

Fig. 13. Shape of enzyme thermal inactivation curves.

from Arrhenius plots are very high, 50 to 150 kcal/mol. This also happens to be the range of energy of activation for thermal denaturation of proteins, when solubility is used as the criterion for denaturation.

At the temperature range commonly used for the thermal inactivation of enzymes (70 to 100° C), such values of activation energy would correspond to Q_{10} values between 10 and 100. In practice, however, it is often found that the inactivation of some enzymes in foods is accelerated only by a factor of 5 or even as low as 2 (peroxidase of sweet corn) when the temperature is raised by 10° C. The amount of information available on the kinetics of enzyme inactivation, either in model system or in actual foods, is too meager to permit a critical explanation for such discrepancies.

Thermal inactivation processes in industry are designed, by and large, empirically.

It must be noted that the resistance of enzymes to thermal inactivation is affected by the composition of their environment. Water activity and pH are particularly important.

4-9. THERMODYNAMIC ASPECTS

It was pointed out in Section 4-2 that the basic condition for a reaction to take place was thermodynamic feasibility and that the presence of an enzyme does not in any way alter this condition. It is appropriate to discuss briefly the laws of thermodynamics and their application to biological reactions.

4-9-1. *Definitions.* The first law of thermodynamics, known also as the law of the conservation of energy, states that the sum of all changes of energy in a system is zero. Energy can cross the borders of a system as work or heat. Therefore, the first law may be stated by the following equation:

$$\Delta E = Q - W \tag{13}$$

where ΔE is the increase in internal energy of the system, Q is the heat supplied to the system and W is the work done by the system on its surroundings.

Chemical reactions are usually constant pressure processes and the only type of work involved is associated with changes in volume, i.e. $P\Delta V$. Under such conditions:

$$Q = \Delta E + P\Delta V \tag{14}$$

Since the right-hand side of the equation consists only of properties of the system, Q must be considered as a change in a thermodynamic property (a state function). This property is defined as *enthalpy*, H. Therefore,

$$\Delta H = \Delta E + P\Delta V \tag{15}$$

If heat is evolved in the reaction, ΔH is negative and the reaction is termed *exothermic*; if heat is absorbed by the system, ΔH is positive and the reaction is termed *endothermic*.

As far as the first law is concerned, any process is possible provided that the principle of energy conservation is observed. We know from experience that this is not true. The second law of thermodynamics formalizes these observations. An additional property, associated with the capacity of a system to undergo a spontaneous change is defined as *entropy* S.

Applied to chemical reactions, the second law states that a system can undergo spontaneous change only if its entropy increases as a result of the process. Applied to a heat engine, the second law states that a system which only absorbs heat and transforms it quantitatively to work is not feasible. A certain amount of the heat absorbed must be rejected and this amount is equal to the expression TdS. At constant pressure, the heat absorbed is ΔH and at constant temperature TdS becomes $T\Delta S$. Thus the change in *free energy* ΔF, i.e. the maximum quantity of energy that can be converted into useful work is given by:

$$\Delta F = \Delta H - T\Delta S \quad (16)$$

In any spontaneous change ΔS must be positive, therefore $\Delta F < \Delta H$. While it is possible to measure ΔH directly by means of a calorimeter, it is impossible to do the same for ΔS or ΔF. These are computed with the help of equations relating these values to other measurable properties of the system. When the system reaches equilibrium, its capacity to undergo further spontaneous changes decreases to a minimum, no further change in free energy is possible and therefore, at equilibrium $\Delta F = 0$.

Consider a reaction represented by the following equation:

$$aA + bB \rightleftharpoons cC + dD \quad (17)$$

The change in free energy associated with this process is given by the following relationship:

$$\Delta F = \Delta F^\circ + RT \ln \frac{[C]^c [D]^d}{[A]^a [B]^b} \quad (18)$$

where:

R = gas constant

$[A]$, $[B]$ etc. are the molar concentrations of reactants A, B etc. ΔF° is a constant value, characteristic of the system. It is termed the "*standard free energy change*" of the reaction. It is a more useful concept than ΔF, because while ΔF depends on the concentration of the reactants, ΔF° at a given temperature and pressure is constant for any given chemical reaction. Its value can be computed as follows: when the reaction reaches equilibrium, $\Delta F = 0$ and the concentrations are interrelated by means of the equilibrium constant K:

$$K = \frac{[C]^c\,[D]^d}{[A]^a\,[B]^b} \tag{19}$$

Hence:

$$\Delta F^\circ = -RT \ln K \tag{20}$$

K is determined experimentally by analyzing the equilibrium mixture. Reactions with a negative change in free energy are termed *exergonic.* Reactions with a positive ΔF° are *endergonic.*

4-9-2. *Energetics of biological reactions.* One of the consequences of the second law of thermodynamics is that an endergonic reaction can take place only if the required free energy is supplied simultaneously from another, sufficiently exergonic reaction. The coupling of reactions in this way is the basis of biological energetics. Such coupling requires the presence of a highly reactive system capable of absorbing and storing the energy released by exergonic reactions in order to supply it when needed by energy-requiring changes. The most important biological "*energy bank*" is the ADP–ATP system.

ADP (adenosine di-phosphate) and ATP (adenosine tri-phosphate) are nucleotides widely distributed in nature. Their importance in energy transfer was demonstrated by Lipmann as early as 1939. Their structure is shown below:

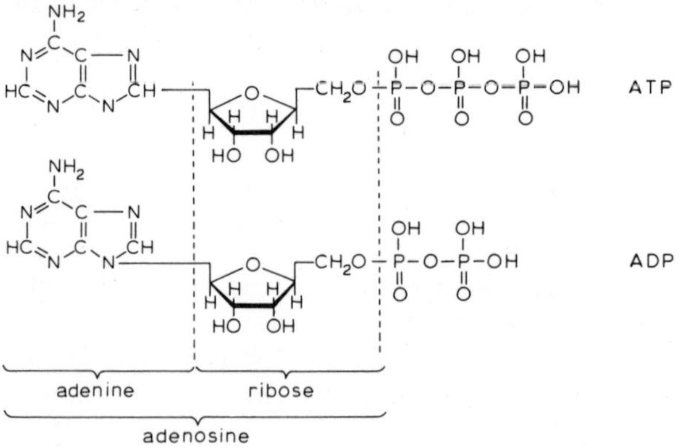

ATP

ADP

adenine

ribose

adenosine

At neutral pH both ADP and ATP occur as highly charged anions, associated usually with bivalent magnesium cations.

In the presence of appropriate enzymes (ATPases), ATP is hydrolyzed into ADP and inorganic phosphate. This is a strongly exergonic reaction with a standard free energy change of —7.3 kcal/mol at pH 7. The reversal of hydrolysis, i.e. the synthesis of ATP from ADP and inorganic phosphate, is obviously an endergonic process with $\Delta F^{\circ} = +7.3$. The system is highly reactive under suitable conditions of ionic dissociation, presence of activators, presence of proteins with ATPase activity. The transfer of energy by means of the ADP–ATP system may be represented schematically as follows:

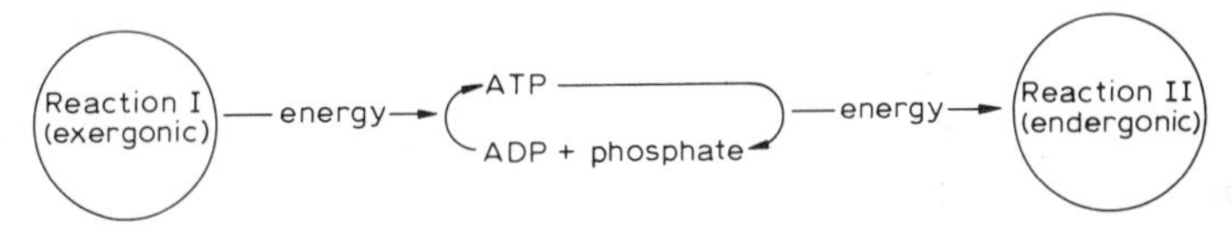

The pyrophosphate bond hydrolyzed in the ATP → ADP process is termed an "*energy rich*" *phosphate bond.* It is schematically represented as ~Ⓟ The second phosphate group is also linked by an energy rich bond, in contrast to the phosphate group nearest to the ribose, which has a ΔF° of hydrolysis of —3 kcal/mol only.

Aspects of energy transfer by means of the ADP–ATP system will be discussed in the next chapter, in connection with muscle contraction. The subject will be further developed in connection with respiration and photosynthesis.

SELECTED BIBLIOGRAPHY TO CHAPTER 4

Acker, L.W., Water Activity and Enzyme Activity, Food Technology, **23** (1969), 1257-1270.

Bull, H.B., An Introduction to Physical Biochemistry, F.A. Davis Co., Philadelphia, Pa., 1964.

Reiner, J.M., Behavior of Enzyme Systems, 2nd edn., Van Nostrand–Reinhold Co., New York, 1969.

Whitaker, J.R., Principles of Enzymology for the Food Sciences, Marcel Dekker Inc., New York, 1972.

CHAPTER 5

PROTEIN SYSTEMS IN FOODS

5-1. PROTEINS OF MEAT AND FISH

Anatomically, meat and the principal edible portion of fish are muscles. To the biologist, muscle is important as the "living engine" where chemical energy is transformed to mechanical work. The food scientist is interested in the eating quality of muscle and its response to processing. It is now clear, however, that many points of close contact exist between the mechanism of movement generation in the living muscle and the behavior of muscle, *post-morten*, as meat. Meat and fish contain between 50 and 80% or more protein on dry basis, this rather large variation being mainly due to the variability of fat content. Qualitatively, proteins are the most important constituent of muscle, whether we consider its biological functions or its characteristics as food.

5-1-1. *The structure and function of muscle.* A muscle consists of fiber-like giant cells, measuring from one millimeter to a few centimeters in length, held together by films of connective tissue. Inside the fibers, bathing in *sacroplasmic fluid* (the cytoplasm of muscle), are found thousands of thread-like bodies, with diameters of about 1 micron, running the length of the fiber. These are named *fibrils*, and constitute the contractile system of the muscle. Electron-microscopic examination of the fibrils reveals the existence of a finer structure consisting of two different types of filaments arranged in hexagonal array (Fig. 14). The thick filaments are molecules of *myosin*, arranged head to tail. The thinner filaments consist of helical molecules of another protein, *actin*. Myosin accounts for about 35% of muscle protein, and has the important feature of being an ATPase. ATPase activity of myosin is situated at the *feet* protruding from the filament (Fig. 15) and connecting between adjacent myosin-actin threads.

Muscle contraction is now explained in the light of the "*sliding filament*" theory. The nervous signal for contraction generates an

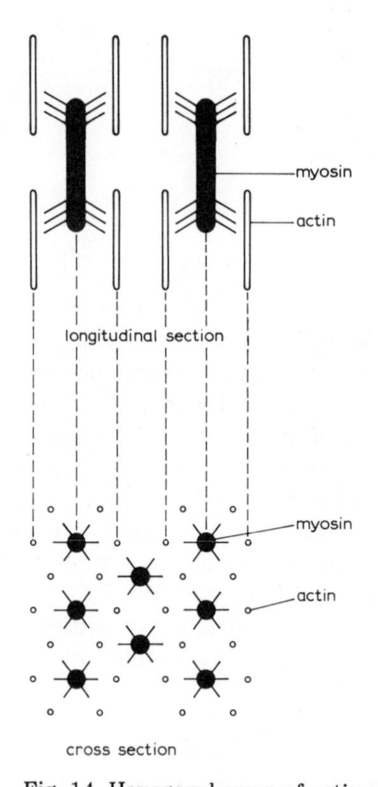

Fig. 14. Hexagonal array of actin and myosin filaments in a muscle fibril.

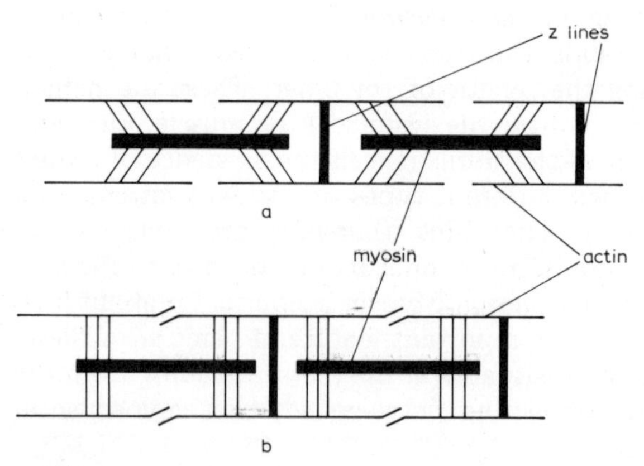

Fig. 15. Sliding filament theory of muscle contraction: (a) relaxed, (b) contracted.

electrochemical impulse causing the release of Ca^{++} ions at the vicinity of the fibrils. ATPase activity of the myosin is enhanced by Ca^{++}. ATP is decomposed and as the concentration of ATP decreases the configuration of the feet is altered as shown in Fig. 15. Actin filaments are pulled one towards the other and the fibril is contracted. The tension produced is transmitted through the connective membranes to the fibers, to the bundles and finally to the entire muscle. Most of the energy liberated from the decomposition of ATP is used as mechanical energy for the generation of work. In the living tissue, ATP is quickly regenerated using the energy of the main fuel system of the body, aerobic oxidation of sugar. In the process of relaxation, Ca^{++} is pumped out from the vicinity of the fibrils, thus de-activating myosin as ATPase. ATP accumulates around the filaments, the feet resume their former angle of recline and the filaments slide back. During relaxation ATP serves as a lubricant between the sliding filaments. Mention should be made of *creatine* which serves as a reserve of high energy phosphate necessary for the regeneration of ATP.

$$\underset{\text{creatine phosphate}}{H_2C{-}N(CH_2{-}COOH){-}C({=}NH){-}NH{\sim}Ⓟ} + ADP \rightleftharpoons \underset{\text{creatine}}{H_2C{-}N(CH_2{-}COOH){-}C({=}NH){-}NH_2} + ATP$$

Creatine is a characteristic constituent of meat. Its concentration in foods is sometimes used as an indication of meat content.

5-1-2. *Rigor mortis.* When an animal dies, the transport of oxygen to its muscle and hence aerobic oxidation of sugar ceases. ATP can no longer be regenerated and the muscle becomes gradually depleted of ATP. The natural lubricant or insulator between actin and myosin filaments is no longer available in sufficient quantity. Cross-linkage occurs between the myosin-actin filaments and the muscle becomes stiff. This condition of stiffness, which occurs shortly after death, is known as *rigor mortis.* Rigor and its side-effects are extremely important in meat technology. Another important change in meat is a marked drop in pH after slaughter, from nearly neutral to acidic (pH 5.6–5.8). This increase in acidity is due to the fact that, at the absence of oxygen only anaerobic glycolysis can occur, the end product of which is lactic acid. Meat in the state of rigor is not only very tough but also less juicy, since for most muscle proteins, isoelectric point and therefore minimum water-binding capacity occurs at pH values near 5.5.

Fortunately the state of stiffness is not permanent. If meat is stored in a refrigerator, it begins to soften after a few days and its water binding capacity increases. In meat technology this process is known as *aging*. The exact nature of the changes which take place during aging is not known; at any rate, this is not believed to be a reversal or resolution of rigor. It is usually assumed that the softening of meat during aging is due to protein hydrolysis, catalyzed by proteolytic enzymes naturally present in the muscle. The experimental evidence in support of this assumption is not conclusive. The extent of proteolysis during aging is very slight. The activity of proteolytic enzymes such as *cathepsins* is much lower in muscle than in organs such as liver and kidney. However, hydrolysis of a few bonds might well be sufficient to release the rigidity of meat. Fish muscle undergoes the same changes but the increase in rigidity after death is less pronounced, especially in white flesh.

5-1-3. *Collagen and gelatin. Collagen* is the main protein constituent of skin, bones and connective tissue. It has a rather unique amino acid composition, with an unusually high content of proline, hydroxyproline and glycine, very little methionine and no cystine or tryptophane. Furthermore, collagen contains some carbohydrate and should be regarded as a glycoprotein. It is a fibrous protein with a considerable degree of crystallinity. The molecules are shaped as rods, consisting of three-strand helices. Collagen is easily denatured by heat. Beef muscle collagen undergoes sudden shrinkage at 64° C, this effect being due presumably to "melting" of the crystalline regions and disorganization of the structure. Collapse of the native structure of collagen is largely responsible for the *shrinkage* of meat during cooking.

Gelatin is the product of thermal dissociation of collagen molecules. In industry, gelatin is prepared by exposing collagen-rich tissue, such as bones or hide, to the prolonged action of hot water with added alkali or acid. Gelatins have molecular weights of about 100,000 as compared to values between 300,000 and over 700,000 obtained for collagen. Upon cooling, aqueous solutions of gelatin solidify. The elastic consistency of *gelatin gels* is very valuable in the manufacture of desserts and confectionery. The gel is thermoreversible, i.e. it "melts" on heating and re-solidifies on cooling. The exact nature of the interchain linkage responsible for the formation of the gel network is still a matter of controversy. It is thought to involve the polypeptide backbone rather than the side chains of the protein. Consolidation of the network may be due to the formation of a few centers of crystallization between the otherwise independent chains.

5-1-4. *Myoglobin, the pigment of red meat.* Color is undoubtedly one of the major quality factors in meat and processed meat products. The characteristic color of red meat comes from a heme-conjugated protein, *myoglobin.* Myoglobin is quite similar to hemoglobin both in structure and in function. It is composed of a globular protein, *globin*, and a prosthetic group consisting of *heme*, a porphyrin ring with an atom of ferrous iron at its center. It has one heme group per molecule while hemoglobin has four. The iron atom is attached to the nitrogens of the four heterocyclic groups of porphyrin. A fifth bond links the iron to a histidine residue of the protein. The remaining sixth site of coordination is occupied by a molecule of water. Just as hemoglobin, myoglobin has a strong affinity for oxygen. Under aerobic conditions, myoglobin adsorbs oxygen reversibly without oxidation of its iron atom. In this process, the water molecule is presumably replaced by O_2. The oxygen-myoglobin complex, named *oxymyoglobin*, has the bright red color characteristic of freshly cut meat. Oxymyoglobin serves as an oxygen reserve in muscles where short periods of intensive work may develop a temporary situation of oxygen deficit. However, oxygen may also cause oxidation of the iron ion from the ferrous to the ferric state. When myoglobin is thus oxidized, a dull brown pigment, named *metmyoglobin*, is obtained. In living tissue, metmyoglobin is quickly transformed back into myoglobin, through the action of reducing substances such as glucose, hence the use (illegal in many countries) of reducing substances such as ascorbic acid and SO_2 for the preservation of the red color in ground meat. Myoglobin undergoes oxidation to metmyoglobin much more readily when globin is denatured, e.g. as a result of cooking. The technological process of *curing* has the objective of preserving the red color in cooked, processed meat products such as sausages and ham.

Basically, the curing process consists of adding to meat small amounts of *nitrites*, and *nitrates.* The main reaction is thought to proceed as follows: in the slightly acidic medium of meat, nitrites yield nitrous acid which is rapidly decomposed:

$$3HNO_2 \rightarrow HNO_3 + 2NO + H_2O$$

Nitric oxide, NO, combines with myoglobin to form a stable red pigment, *nitrosomyoglobin*:

$$\text{globin}—Fe(N_4)—H_2O \underset{}{\overset{NO}{\rightleftharpoons}} \text{globin}—Fe(N_4)—NO$$

myoglobin nitrosomyoglobin

After thermal denaturation of the protein during cooking nitrosomyoglobin is transformed to *nitrosohemochrome*, a relatively stable pigment (probably dissociated from the denatured protein) which imparts to processed meats their desirable pink color.

Nitrates are not directly active in the process of color fixation, but they are first transformed to nitrite under the reducing conditions prevailing in cured meat. The use of nitrates is based on safety considerations since nitrites are toxic and a slight excess of uncombined nitrite would be harmful. Reducing substances (sugars, ascorbic acid) or microorganisms capable of reducing NO_3^- to NO_2^- (such as lactic acid bacteria) are usually added to curing agents. Lately, the safety of using nitrites and nitrates in foods has been questioned. Under certain conditions nitrite is known to react with secondary amines to form *N-nitrosamines* as follows:

$$R_1R_2NH + HNO_2 \qquad R_1R_2N{-}NO + H_2O$$

sec. amine N-nitrosamine

Nitrosamines are suspected to cause cancer. Some substances with secondary amine structure (e.g. proline) are present in foods and others could be formed, at least theoretically, when food is cooked. The importance of the nitrosamine problem in practice is still a highly controversial subject.

5-2. THE PROTEINS OF MILK

The function of milk in nature is to provide the new born mammal with his first nourishment and to transfer into the young body some of the immunologically active substances necessary for its protection during the first days of life. Milk proteins have been investigated extensively and many breakthrough achievements have been gained, especially in the last decade, towards better understanding of this complex system. The composition of milk varies within wide limits from one mammal to another. In the following discussion, "milk" will be used to designate cow's milk since this is the material of principal interest to the food technologist.

Physically, milk consists of three distinct phases: a continuous aqueous solution (*milk serum*), and dispersed in it, very small *fat globules* and even smaller solid particles (micelles) of *casein*. Proteins are present and important in all these three phases.

Caseins: Quantitatively this is the most important group of milk proteins, accounting for over three quarters of nitrogen in milk. *Casein* occurs in milk as colloidal particles with an average diameter of 1200 Å. It is a conjugated protein containing phosphoric acid, calcium and some carbohydrate. Phosphoric acid is found primarily as a monoester of the serine residues. Casein is easily isolated by isoelectric precipitation at pH 4.6.

The heterogeneity of casein was shown by Linderstrøm-Lang and Kodama already in 1925. Refinement of ultracentrifugal analysis permitted fractionation of whole casein to various components. Using the then (1939) novel technique of electrophoresis Millander was able to separate casein into three electrophoretic fractions which he named α, β and γ caseins. These three components account very roughly for 65, 30 and 5% of the total casein, respectively. In 1955, von Hippel and Waugh demonstrated that α-casein consists of two components which they designated as α_s-casein and κ-casein. These two components differ in the solubility of their calcium salts, calcium α_s-caseinate is insoluble while calcium κ-caseinate is soluble. The importance of this discovery in our understanding of micelle structure and milk curdling will be discussed later. More recently, even these fractions, as well as β and γ caseins, were shown to consist each of several components.

All three major casein groups, namely α_s, β and κ-caseins, have molecular weights near 25,000. They differ considerably in their phosphorus content, which is approximately 1% in α_s, 0.6% in β and 0.2% in κ-casein. Only κ-casein contains carbohydrate.

Micelle structure and coagulation

The white, opalescent appearance of milk is due to the presence of organized aggregates, or *micelles*, of casein. Casein micelles are spherical bodies, with a diameter range of 400 to 2800 Å (1200 Å average). The number of micelles present in 1 ml of milk is in the order of 10^{13}. The exact structure of casein micelles is still a subject of active investigation. A number of models have been suggested, accepted for a while and then rejected in the light of new experimental evidence.

Recently Waugh and Noble proposed a model which seems to reconcile a good deal of observations. According to this model the casein micelle would have a *core* consisting mainly of α_s and β Ca-caseinates and a *coat* of Ca κ-caseinate. κ-casein serves as the protecting colloid of the micelle. In its absence, calcium complexes of other caseins would precipitate.

In the manufacture of cheese, preparations obtained from the stomach of the calf are used for the coagulation of milk. The active principle of these preparations is the enzyme *rennin* (also known as *chymosin*). The mechanism of rennin coagulation of casein has been one of the most puzzling problems in the biochemistry of milk proteins. The overall reaction has been thought to involve some ill-defined alteration of casein and its transformation to *paracasein*, giving insoluble calcium-paracaseinates (curd) in the presence of Ca^{++}. Rennin is a protease but its maximum proteolytic activity is observed at pH 3.8. In cheese manufacture milk is clotted at pH 5-5.5. The proteolytic activity of rennin under these conditions is negligible. Thus, when rennin causes curdling of milk, its proteolytic action is necessarily limited and probably specific. Furthermore, α_s and β-caseins are not changed in the process. Only κ-casein is affected. A portion of the κ-casein molecule is split into large peptides (macropeptides), some containing carbohydrate. The residual κ-casein, reduced in size and free of carbohydrate, is named *para κ-casein.* Thus, rennin does not clot milk directly but destroys the protective coat of the micelles by catalyzing the partial hydrolysis of κ-casein, at some particularly labile bond. Deprived of the protective colloid, the calcium complexes of caseins undergo coagulation and clotting occurs.

Other proteolytic enzymes such as pepsin and papain also cause rapid clotting of milk but the properties of the curd are probably not the same.

Precipitation at the isoelectric pH of 4.6 has been the conventional procedure for the isolation of casein from milk. Curdling by acidification is also practiced in milk technology, for the preparation of sour milk products and some soft cheeses. "Natural" acidification is achieved by inoculating milk with starters, i.e. cultures of lactic acid bacteria. These microorganisms transform the lactose into lactic acid. When the pH approaches the isoelectric value, the viscosity increases and finally very thick or gel-like products, such as yogurt, are obtained. The conditions necessary for the formation of the gel establish a delicate balance between participation and hydration. Complete curdling can be achieved by heating the gel.

In all types of milk curdling, the presence of Ca^{++} is necessary. In milk, Ca^{++} is partly bound to the protein through acidic functional groups (carboxyl and phosphate monoester) and partly in combination with phosphoric acid as colloidal calcium phosphate.

Milk serum proteins

The non-casein proteins of milk are also designated as "*whey proteins*" because they are not affected when casein is precipitated during the curdling process and remain in solution in the supernatant (whey). Various separation techniques reveal the presence of numerous whey proteins, the principal being *α-lactalbumin* (0.7 g/liter), *β-lactoglobulin* (3 g/liter), *immunoglobulin* (0.6 g/liter), *bovine serum albumin* (0.3 g/liter). Some whey proteins and particularly globulins are antibodies important in the transfer of immunity from mother to offspring. Some minor whey proteins are enzymes.

Fat globule membrane proteins

A relatively small amount of protein (less than 1 g per 100 g of fat) is adsorbed in the colloidal film surrounding the milk fat globules. This film is thought to control the stability of the fat-in-water dispersion in milk, but it is also found to contain a number of enzymes. The fat globule is coated with membrane material as it leaves the lactating cell. It is believed that the origin of the fat globule membrane is in the wall membrane of the lactating cell itself. Fat globule membrane proteins are phospholipoproteins containing also some carbohydrate. The physical structure of the membrane is not well understood.

5-3. THE PROTEINS OF EGG

Egg white proteins differ markedly from the proteins of egg yolk in biological function and in composition. The technological functions usually assigned to these two distinct portions of the egg are also different. While the most important functional property of egg white in food products is its ability to form stable foams, the functional significance of egg yolk is largely connected with its capacity to stabilize fat-water emulsions.

For obvious reasons we shall limit our discussion to avian egg.

Egg white is almost a pure solution of protein. The German word for proteins is "*Eiweiskörper*", literally meaning "egg-white substances". Almost all the proteins of egg white are biologically active as enzymes, inhibitors or antibodies. The main function of egg white in nature is to protect the embryo against microbial factors. The proteins of egg white are numerous. The most abundant is *ovalbumin* (54%), followed by *conalbumin* (13%), *ovomucoid* (11%), *lysozyme* (3.5%), and others. Ovalbumin has a molecular weight of 45,000, is easily denatured and has a high content of reactive —SH groups. Conalbumin is one of the antibacterial proteins of egg white. It has the interesting ability to bind iron and other metal ions. This property is believed to be the reason for its antibacterial activity against microorganisms requiring iron for growth. Ovomucoid is a glycoprotein containing approximately 20% carbohydrate, mostly hexosamines. It inhibits the proteolytic activity of trypsin. Lysozyme is an enzyme. It causes *lysis* (dissolution) of certain microorganisms by breaking down carbohydrate polymers found in their cell walls.

The antibacterial function of egg-white proteins is of consequence in the use of eggs as food. The unusual resistance of eggs (as compared to milk or meat) to spoilage is due to the presence of an efficient protective system consisting of a physical shield (egg shell and its membrane) and a chemical barrier (the antimicrobial proteins of egg white).

Egg white proteins are easily denatured. The foaming capacity may be impaired or lost completely as a result of thermal processing, freezing, dehydration or even mechanical treatment involving high shear rates. Methods have been developed to minimize processing damage to egg-white proteins.

Egg yolk proteins are practically devoid of biological activity and serve mainly as a food reserve for the young chick. The main components are two lipoproteins (*lipovitelin* and *lipovitellinin*), a phosphoprotein (*phosphovitin*) and a water soluble fraction (*livetin*).

Egg yolk is used extensively as an emulsifier in mayonnaise, cake batters and confectionery. The ability of egg yolk to stabilize fat-water dispersions rich in fat may be assigned partly to lipoproteins and partly to uncombined phospholipids.

5-4. SEED PROTEINS

The universal importance of *cereal grains* as a source of food is obvious. *Legume seeds* such as beans, peas, chickpeas, lentils are valuable constituents of the human diet in many parts of the world. Modern production of milk, meat and eggs depends largely on the availability of *oilseeds* or oilseed meals which represent the main source of protein in animal feed mixtures. In comparison to other plant tissues, seeds are relatively rich in protein. The biological role of proteins in the seed is still obscure. The process of germination involves intense and many-sided biochemical activity, requiring the rapid biosynthesis of many enzymes. One can assume that seed proteins provide both the machinery and the raw materials for such biosynthesis.

It is also logical to suppose that, as most other constituents of the cotyledons, seed proteins serve as a food reserve for the seedling, providing the young plant with amino acids and nitrogen until the root system and the photosynthetic apparatus are sufficiently developed. Within the cells of the seed cotyledons these "storage proteins" occur in granules with diameters in the range of 2-20 μ, known as *aleurons* or "*protein-bodies*".

Interest in seed proteins arose early in the history of protein chemistry. As Altschul points out, the first enzyme to be crystallized, *urease*, was isolated from a plant seed. Nevertheless, progress in the basic biochemistry of seed proteins has been slow.

5-4-1. *Cereal proteins.* The protein content of cereal grains is in the vicinity of 10% (wheat and barley 13%, rice and maize 9%). Space does not permit a description of the protein systems in all important cereals. In the following discussion we shall concentrate on wheat proteins.

As early as 1745 Beccari isolated *gluten*, the main protein fraction of wheat. Gluten is the yellowish, elastic material which remains after wheat flour dough has been kneaded under a stream of water to remove starch and water soluble constituents. Later, Osborne (1907) classified wheat proteins in five groups according to solubility.

According to his classification, gluten consists of two components: *Gliadin* (soluble in alcohol) and *glutelin* (insoluble in alcohol). In the water-soluble fraction Osborne characterized an albumin, a globulin and a fifth, ill-defined component named "proteose".

Gluten is the principal protein of wheat, accounting for nearly 80% of the nitrogen. It is the factor responsible for the unique ability of wheat flour to form the well-known leavened bread texture. Gluten makes possible the formation of a dough, which can retain the carbon dioxide produced during fermentation or liberated by chemical leavening agents. When water is added to wheat flour and the dough is kneaded, gluten particles swell considerably, aggregate, and form a continuous phase with peculiar rheological properties. A good dough is sufficiently plastic to permit shaping and increase of loaf volume as a result of gas generation. However, the dough must possess also sufficient mechanical strength and elasticity in order to conserve its form, volume and its well-known porous "comb texture". Too much strength would oppose excessive resistance to volume increase and would result in loss of the ability of dough to retain gas during baking. The rheological properties of wheat gluten are therefore of great technological importance. Unfortunately, interpretation of the viscoelastic properties on the basis of gluten structure at molecular level is difficult.

Chemically, gluten is now known to contain many more components than the two fractions defined by Osborne. Several "gliadins" have been isolated by chromatographic fractionation techniques. Glutelin usually appears as a single component with a molecular weight in the order of 10^6, but it is suspected to consist of many smaller units cross-linked through S—S bonds. These bonds are responsible for much of the dough's elasticity and mechanical strength. When prepared by washing of dough, crude gluten contains considerable quantities of lipids. A large part of these lipids combine with the protein when water is added to flour but some constituents of gluten are presumably lipoproteins in the native state. Glutelins seem to combine with lipids more readily than do gliadins. Recently, a model has been proposed, picturing the continuous gluten phase of wheat dough as a sheet structure, composed of flat protein "platelets", sandwiching with layers of lipoproteins (Fig. 16). The lipid would act as a lubricant when the sheet is deformed. The improvement in baking quality observed when fats and phospholipids (lecithin) are added to dough seems to confirm this view.

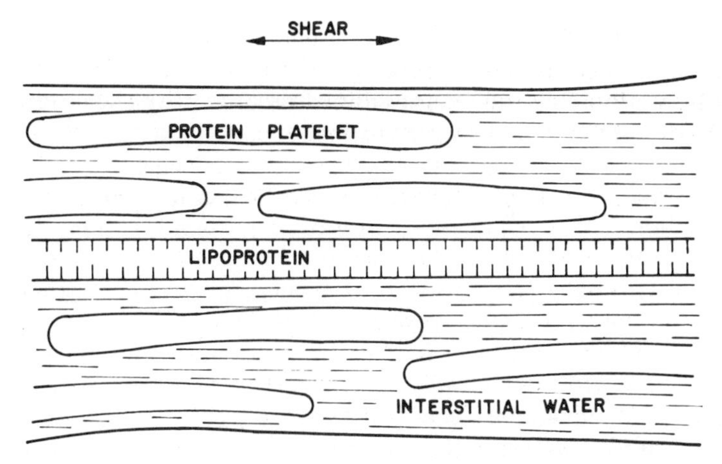

Fig. 16. Hypothetical model of gluten sheet (From Wheat Chemistry and Technology, I. Hlynka, Ed.).

The technological significance of water-soluble wheat proteins, other than some enzymes, has not been elucidated. Finally, mention should be made of an interesting substance named *purothionin* and crystallized by Balls, Hale and Harris from petroleum ether extracts of flour. With a molecular weight of about 10,000, this substance seems to be something between a small protein and a large peptide, in combination with considerable amounts of lipid. Its activity as a fermentation inhibitor factor has been investigated.

Quantitatively, cereal grains supply a substantial part of the protein in the diet, especially in populations relying on bread or rice as the main staple. Qualitatively, cereal proteins are of rather low nutritional value by virtue of their low content in some essential amino acids, mainly lysine. The value of cereals as a source of protein can be increased efficiently and economically by supplementation with other proteins richer in lysine (milk, fish protein concentrate, legumes, soybean flour) or with the limiting amino acid itself.

5-4-2. *Oilseed proteins.* Soybeans, peanuts, cottonseed, sunflower, rape and a number of other seeds, grown mainly as sources of edible oil, are named "*oilseeds*". Most of these seeds have, in addition to oil, a high content of protein. Oil cakes or "*meals*", which are the material left after the oil is extracted, are one of the major sources of protein in the world. At present, oilseed meals are used almost exclusively in animal feeds but recently, much work has been done for the transformation of these sources into foods fit for human consumption. At the same time considerable research effort has been devoted to the characterization and biochemical study of these proteins.

Oilseeds belong to different botanical families and there is no justification, other than convenience, for grouping them together. As in the case of other seeds, gross classification of the proteins on the basis of solubility shows the presence of a major globulin fraction. More discriminating techniques reveal that fractions which were thought to be homogeneous are, in fact, mixtures of a number of components. These are now usually designated by their sedimentation constants. For example, the principal protein of soybeans, characterized as a globulin and named *glycinine* by Osborne, is now known to contain several, separable components. The major globulin of peanuts consists, in fact, of several sub-fractions named arachin and the conarachins.

While the major storage proteins of oilseeds are of obvious importance in nutrition, it is rather the minor components that have been attracting most of the attention in recent oilseed protein research. Many of these components have toxic properties. The best known examples are the *trypsin inhibitors* of soybeans (Section 3-7). Trypsin inhibitors are also present in peanuts, sunflower and other oilseeds. The same oilseeds as well as rapeseed, and especially castorbean meal, contain proteins which have the power of agglutinating red blood cells (*hemagglutinins*). Most of these undesirable components can be inactivated by proper heat treatment (toasting) or eliminated by separation. Protein "refining" processes are commercially applied to soybean and cottonseed meals. *Soybean meal* contains 44-50% protein. The physiologically harmful proteins (trypsin inhibitors, hemagglutinins) are relatively small, soluble polypeptides. In addition, soybeans contain peculiar sugars of difficult digestion such as *stachiose* (Section 7-3). If soybean meal is extracted with aqueous alcohol, the sugars and most of the smaller polypeptides are removed. The remaining material containing 70% protein is produced and sold as a "*soybean protein concentrate*". Another method of purification consists of extraction of the protein in water or dilute alkali, followed by precipitation at pH 4-4.5. The washed precipitate (curd, *isolated soybean protein*) is almost pure protein, free of trypsin inhibitors or hemagglutinins, which remain in solution at this pH range. Over two thirds of the meal protein is recovered as isolate.

An important development towards the increased used of oilseed proteins as human foods is the advent of processes known as *texturization* or "*structuring*" whereby a desired texture (mainly simulated meat texture) is imparted to the material. In one of such processes, an alkaline solution of isolated soybean protein is forced through narrow nozzles into a coagulating acid bath. The stream coagulates in the form of a fiber which is immediately stretched and spun, much in the same way as synthetic fibers are produced for the textile industry. The soybean protein fibers are bound together by means of a suitable flavored gel to produce a meat-like structure.

In cottonseed, the main problem is the presence of a phenolic substance, *gossypol*, which is objectionable not only as pigment but also for its toxic properties. Gossypol and its removal will be discussed in Chapter 16.

5-5. SINGLE CELL PROTEIN

The name "*Single Cell Protein*" (SCP) came into existence on the occasion of the first international conference on the subject at the Massachusetts Institute of Technology, in 1967. However, the use of monocellular organisms as a source of protein is much older.

Yeast of the Saccharomyces species (baker's yeast and brewer's yeast) has been long used as a minor food ingredient, probably for its peculiar flavor and high content of vitamins of the B group. During World War I, another type of yeast, Torula (Candida) utilis was produced in vast quantities in Germany as a protein source for use in animal feeds. Because of its less exacting nutrient requirements Candida is more economical to produce. Interest in single cell protein arose considerably in the recent years, as a result of the forecasted shortage of conventional protein sources. *Bacteria*, *fungi*, *algae* as well as *yeast* have been considered.

Industrially, the production of single cell protein may be represented with the help of the following diagram:

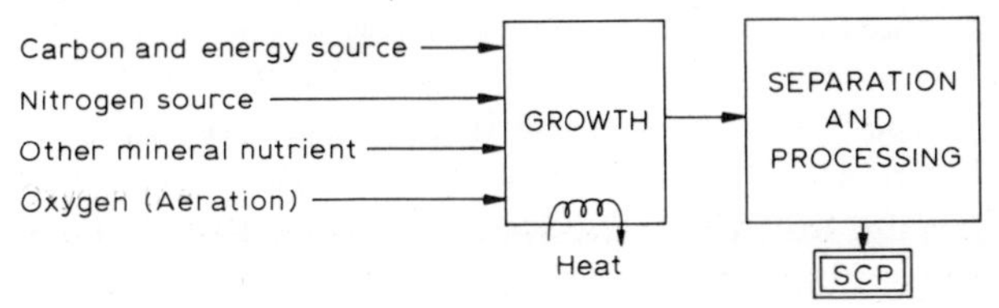

Carbohydrates from agricultural wastes and by products (molasses, sulfite, liquor from paper industry, waste starch, citrus peel liquor, etc.) and recently hydrocarbons (crude petroleum, *n*-paraffin fractions) or partially oxidized organic substances (methanol) are the main sources of carbon and energy. The microorganism obtains the energy necessary for growth from oxidation of the carbon source. Oxygen is essential for the process and is supplied by aeration. Aeration is one of the major cost factors in the production of SCP. Excess heat must be removed to keep the temperature in the growth vessel (fermentor) within acceptable limits. Heat removal is another important cost factor. The source of nitrogen is usually ammonia or ammonium salts.

In the case of green algae and other microorganisms capable of photosynthesis, sunlight energy and CO_2 from the air are utilized. The process then becomes a complete simulation of agriculture in a fermentor, without the use of land. Unfortunately, economical production of SCP from such microorganisms has not been possible.

Separation of the cells from the culture medium is another problem, especially in the case of bacteria. Centrifuges are commonly used for this purpose. Removal of residues carried over from the growth medium is essential, especially when the organism is grown on hydrocarbon. Further processing may include disintegration of cell walls, autolysis, removal of nucleic acids and finally drying.

Nutritionally, single cell proteins show fairly adequate amino acid composition. Sulphur containing amino acids are in short supply in yeast and most fungi, but adequate in several bacteria considered for commercial production. The high content of nucleic acids (see Section 6-4-4) is a problem in human nutrition. The overall wholesomeness of the products as human foods has to be evaluated carefully, especially in the case of microorganisms grown on petroleum material. At present, despite the growing interest in the subject, the use of single cell protein is still limited to animal feeding.

5-6. LEAF PROTEINS

Green leaves are the site of tremendous metabolic activity in plants and one can expect to find there a large supply of proteins, needed for the development of enzymic systems. In fact, green plants — especially grasses — are an important source of protein in animal husbandry. One of the major green crops used as fodder, alfalfa, contains typically 18% protein on dry basis. Leaves contain more proteins when they are young.

The use of leaves as a significant source of protein for humans, beyond the conventional consumption as leaf vegetables, is difficult. Leaves contain a high proportion of undigestible fiber (cellulose, lignin, etc.), often two or three times higher than the protein content. The taste and color are often objectionable. The protein fraction can be concentrated or isolated but the acceptability of the concentrates is still poor and the economics of the process are not attractive.

5-7. PROTEOLYTIC ENZYMES IN FOOD PROCESSING

The use of proteases (peptidases) in food industries was mentioned on several occasions. Additional applications will now be summarized.

(a) *Tenderization of meat.* The natural tenderization of meat after rigor mortis involves proteolysis with the help of cathepsins (Section 5-1). It can be accelerated by the intentional addition of proteolytic enzymes. For the purpose of tenderization, it is not necessary nor desirable to bring about extensive proteolysis. Liberation of peptides and amino acids would impair the flavor of meat. Obviously only endo-peptidases are suitable as tenderizers. The most commonly used protease is *papain,* followed by *bromelain* and *ficin.* The methods of application include *pre-slaughter injection, post-slaughter injection* and *surface application.* Tenderizing enzymes do not only affect the myosin–actin system but they also hydrolyse collagen and elastin and thus soften the connective membranes as well.

(b) *Baking industry.* Fungal proteases are sometimes used in the bread baking industry. Slight, controlled proteolysis of wheat gluten lowers the dough viscosity without impairing the mechanical strength necessary for gas retention. The reduction in viscosity shortens mixing time and is important in modern bakery practices using high-speed mixers. The texture of the final product is somewhat improved.

(c) *Chill proofing of beer.* When beer is chilled to temperatures below 10° C a fine precipitate known as the "*chill haze*" occurs. The precipitate is apparently a protein–tannin complex, insoluble in the cold beer. The addition of a protease which can survive the heat treatment given to beer (pasteurization) solves the problem. Papain is most commonly used for this purpose.

SELECTED BIBLIOGRAPHY TO CHAPTER 5

Altschul, A.M. (Ed.), Processed Plant Protein Foodstuffs, Academic Press, New York, 1958.

Altschul, A.M., Neucere, N.J., Woodham, A.A. and Dechary, A.A., A New Classification of Seed Proteins, Nature, **203** (1964), 501.

Briskey, E.J., Cassens, R.G. and Trautman, J.C. (Eds.), The Physiology and Biochemistry of Muscle as Food, The University of Wisconsin Press, Madison, 1966.

Dyer, W.J. and Dingle, J.R., Fish Proteins with Special Reference to Freezing, in Fish as Food, G. Borgstrom (Ed.), Academic Press, New York, 1961.

Fevold, H.L., Egg Proteins, Adv. in Protein Chemistry **6** (1951), 201.

Lijinsky, W. and Epstein, S.S., Nitrosamines as Environmental Carcinogens, Nature, **222** (1970) 21.

Mateles, R.I. and Tannenbaum, S.R. (Eds.), Single-Cell Protein, M.I.T. Press, Cambridge, Mass., 1968.

McKenzie, H.A. (Ed.), Milk Proteins, 2 Vols., Academic Press, New York, 1970-1071.

Osborne, T.B., The Proteins of the Wheat Kernel, Carnegie Inst. Wash., Pub. No. 84, 1907.

Pomeranz, Y. (Ed.), Wheat, Chemistry and Technology, American Association of Cereal Chemists Inc., St. Paul, Minn., 1971.

Price, J.F. and Schweigert, B.S. (Eds.), The Science of Meat and Meat Products, W.H. Freeman and Co., San Francisco, 1971.

Schultz, H.W. and Anglemier, A.F. (Eds.), Proteins and their Reactions, The Avi Publishing Co. Inc., Westport, Conn., 1964.

CHAPTER 6

CARBOHYDRATES — MONOSACCHARIDES

6-1. DEFINITION, OCCURRENCE, CLASSIFICATION

It is impossible to give an exact definition of the term "*carbohydrate*". Early in the history of organic chemistry, it was observed that a large number of natural substances such as sugars, starch and cellulose, all have empirical formulae of the type $C_x(H_2O)_y$ and can be regarded, at least stoichiometrically, as "hydrates of carbon", hence the name "carbohydrates". It was soon found that these substances are not all "hydrates" and that the stoichiometrical generalization is not always valid. However, the name "carbohydrates" was retained to designate this large and important group of natural compounds. Chemically, carbohydrates are *polyhydroxy aliphatic compounds bearing also carbonyl or carboxyl groups* and their simple derivatives.

These compounds, in one form or another, constitute more than one-half of the organic matter upon the Earth: the greatest part of plants is built of carbohydrates while the animal world contains rather limited amounts of them.

Various sugars, starches, cellulose, hemicelluloses, pectins, numerous gums and mucilages are all carbohydrates. The simple sugars, glucose and fructose, are found in honey and various fruits. These are often accompanied by combined sugars, the disaccharides, of which sucrose, for instance, found in sugar beets and in sugar cane, is the most representative example. More complicated carbohydrates are polymers of the simple sugars or of their derivatives and serve as the "building bricks" in plants (such as cellulose, hemicellulose and pectins) or as reserve materials for future use (such as starches) or probably also as waste products, such as the numerous gums produced by plants. Glycogen, produced in the liver, is also a carbohydrate polymer, constituting a small energy reserve in animals. All these are only a few examples, but the number of carbohydrates in nature is indeed prodigious.

Carbohydrates which cannot be broken down by hydrolysis into simpler, smaller ones are termed *monosaccharides.* The monosaccharides, or simple sugars, are neutral, crystallizable and diffusible substances, readily soluble in water, soluble with difficulty in alcohol and insoluble in ether. Not all of these "sugars" are sweet; they cover a wide range of sweetness and some are even bitter.

Depending on the position of the carbonyl group in the molecule, monosaccharides may be polyhydroxyaldehydes (aldoses) or polyhydroxy ketones (ketoses).

glyceraldehyde (aldose)	dihydroxyacetone (ketose)
CHO	CH_2OH
CHOH	CO
CH_2OH	CH_2OH

Depending on the number of carbon atoms in the molecule, monosaccharides are termed bioses (C_2), trioses (C_3), tetroses (C_4), pentoses (C_5), hexoses (C_6), heptoses (C_7) etc.

The monosaccharides of primary importance are hexoses and pentoses.

6-2. STRUCTURE

Of the many hexoses which are possible on a theoretical basis, only five have been found *in the free state* in the plant kingdom, three of them aldoses (*glucose*, *mannose* and *galactose*) and two ketoses (*fructose* and *sorbose*):

	D-glucose	D-mannose	D-galactose	D-fructose	L-sorbose
1	CHO	CHO	CHO	CH_2OH	CH_2OH
2	HCOH	HOCH	HCOH	C=O	C=O
3	HOCH	HOCH	HOCH	HOCH	HOCH
4	HCOH	HCOH	HOCH	HCOH	HCOH
5	HCOH	HCOH	HCOH	HCOH	HOCH
6	CH_2OH	CH_2OH	CH_2OH	CH_2OH	CH_2OH

The hydroxyl of carbon atom 3 in all these sugars is oriented in the same direction with regards to the whole molecule (in the above formulae to the left). Mannose is different from glucose in its configuration at carbon atom 2 and is therefore called an epimer of glucose. Sorbose differs from fructose in its configuration at carbon atom 5.

Monosaccharides crystallize with relative difficulty and, therefore, Emil Fischer, who truly deserves to be called the father of carbohydrate chemistry, introduced *phenylhydrazine* $C_6H_5NH{\cdot}NH_2$ as a reagent which combines easily with the sugars giving beautiful yellow crystals of osazones. This is a most useful means for the characterization of monosaccharides. The following scheme shows the introduction of two molecules of phenylhydrazine from the reaction mixture. Glucose, mannose and fructose give identical osazones because of the similarity of their configurations at carbon atoms 4, 5 and 6. Hydrazones of the aldoses and ketoses are different in their properties.

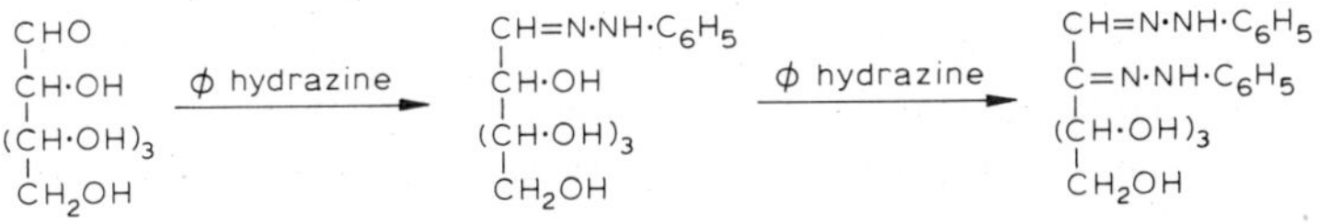

$$\begin{array}{l} CHO \\ | \\ CH{\cdot}OH \\ | \\ (CH{\cdot}OH)_3 \\ | \\ CH_2OH \end{array} \xrightarrow{\phi\ \text{hydrazine}} \begin{array}{l} CH{=}N{\cdot}NH{\cdot}C_6H_5 \\ | \\ CH{\cdot}OH \\ | \\ (CH{\cdot}OH)_3 \\ | \\ CH_2OH \end{array} \xrightarrow{\phi\ \text{hydrazine}} \begin{array}{l} CH{=}N{\cdot}NH{\cdot}C_6H_5 \\ | \\ C{=}N{\cdot}NH{\cdot}C_6H_5 \\ | \\ (CH{\cdot}OH)_3 \\ | \\ CH_2OH \end{array}$$

glucose — glucose-phenylhydrazone — glucose-phenylosazone

The generalized open chain formula for a hexo-aldose shows the presence of four asymmetric carbon atoms (indicated by asterisks).

$$\overset{6}{CH_2OH}{\cdot}\underset{*}{\overset{5}{C}}HOH{\cdot}\underset{*}{\overset{4}{C}}HOH{\cdot}\underset{*}{\overset{3}{C}}HOH{\cdot}\underset{*}{\overset{2}{C}}HOH{\cdot}\overset{1}{CHO}$$

Consequently, the possibility of $2^N = 2^4 = 16$ stereoisomers (8 antipodal pairs) exists. Glucose is only one of these 16 isomers and it should exist in two enantiomorphs, the D- and L-configurations. At this point it is appropriate to mention that in the nomenclature of sugars, it is customary to designate the enantiomorphs by majuscule letters D and L. These do not indicate the direction of the optical rotation of the sugars, but the configuration of the hydroxyl on the carbon atom before the last in relation to that of the last one. *Glyceric aldehyde* is taken as an example:

$$\begin{array}{c} CHO \\ | \\ HCOH \\ | \\ CH_2OH \end{array} \qquad\qquad \begin{array}{c} CHO \\ | \\ HOCH \\ | \\ CH_2OH \end{array}$$

D-glyceric aldehyde — L-glyceric aldehyde

All other sugars are related to these designations of glyceric aldehyde. In order to designate the direction of optical rotations of sugars the signs (+) and (−) are used. Thus D(+) glucose shows that this particular glucose has the D-configuration and an optical rotation to the right, while D(—) fructose designates that this sugar has the same D-configuration but it is levorotatory.

The open chain formula postulated by Fisher was soon found to be inadequate. Although sugars behave somewhat as carbonyl compounds their reactions are not as characteristic as those of true aldehydes or ketones. It is known that aldehydes tend to condense with alcohols to form *hemiacetals*:

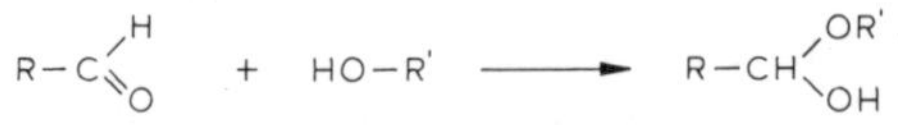

$$R-C\begin{smallmatrix}H\\ \\ O\end{smallmatrix} + HO-R' \longrightarrow R-CH\begin{smallmatrix}OR'\\ \\ OH\end{smallmatrix}$$

According to the open chain formula both the carbonyl and hydroxyl groups necessary for the formation of a hemiacetal are present in the same molecule. The absence of typical aldehydic properties can be explained by the formation of an *inner hemiacetal bridge*, and therefore a closed ring structure

$$\begin{array}{l} HC=O \\ \vdots \\ HC-OH \\ \ \ | \end{array} \longrightarrow \begin{array}{l} \ \ \ \ OH \\ \ \ \ \ \ | \\ H-C- \\ \ \ \ \ \vdots \ \ \ \ O \\ H-C- \\ \ \ \ \ \ | \end{array}$$

With the formation of the hemiacetal ring, the carbonylic carbon (C_1) becomes asymmetric and therefore two stereoisomers are possible around this position.

$$\begin{array}{c} H \ \ \ OH \\ C \\ | \ \ \ \ O \\ \ \ \ \ \ | \end{array} \qquad \begin{array}{c} HO \ \ \ H \\ C \\ | \ \ \ \ O \\ \ \ \ \ \ | \end{array}$$

The existence of two glucose stereoisomers was demonstrated by Tanret. A freshly prepared solution of glucose has a specific rotation of $+113°$, which soon begins to decrease and drops finally to $+52.7°$. This phenomenon, discovered in 1847, is termed *muta-rotation*. In 1895 Tanret succeeded in crystallizing glucose from a solution at 98°C. This glucose had an initial specific rotation of $+19°$, which increased slowly to $+52.7°$. Tanret's glucose received the notation of *β-glucose*, while ordinary glucose was named *α-glucose*. Mutarotation and other observations indicate that these two isomers are easily interconvertible through the open chain structure. Glucose in solution is a mixture of α and β configurations in tautomeric equilibrium.

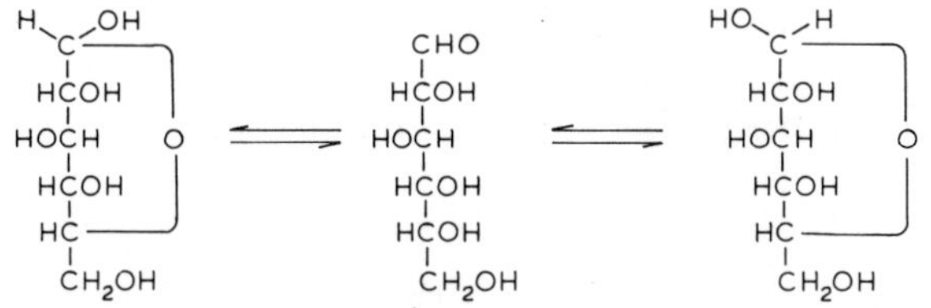

α-D(+) glucose $[\alpha]_D = +113°$ ⇌ open-chain glucose ⇌ β-D(+) glucose $[\alpha]_D = +19°$

Most of the monosaccharides and some of the disaccharides undergo mutarotation. Mutarotation is catalyzed by the addition of H^+ or hydroxyl (OH^-) ions. An enzyme *mutarotase*, present in extracts from the mold *Penicillium notatum* and from some animal tissues (kidney and liver), also catalyzes the mutarotation of glucose.

With the acceptance of the ring structure for monosaccharides the question arose: what is the size of this ring; for, in fact, one can imagine the linkage of carbon atom 1 with any one of the other carbon atoms.

Considerations of bond strain as well as direct experimental evidence indicate that the most stable configurations are the six-member or the five-member heterocyclic rings, similar to *pyrane* and *furane*, respectively.

pyrane 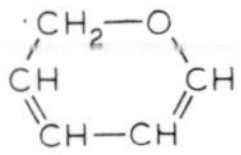 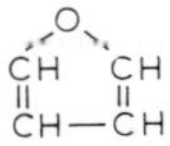 furane

The sugars having a six-member ring are, therefore, named *pyranoses* while those having a five-member ring are called *furanoses.* The pyranose form is the more stable of the two and is, therefore, predominantly found in nature.

The cyclic structure of the monosaccharides is conveniently depicted by the valuable method suggested by Haworth in 1927. As drawn in the following formulae, the plane of the pyranose ring is shown in perspective to be perpendicular to that of the page on which it is written, with the substituents above or below the plane of the ring:

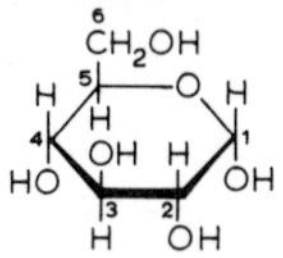

α-D(+) glucopyranose 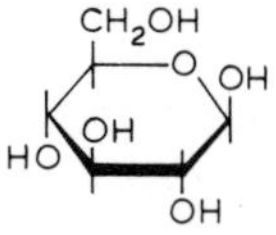 β-D(+) glucopyranose

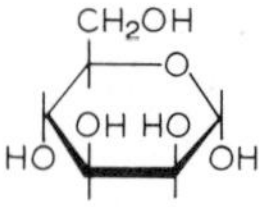

α-D(−) mannopyranose

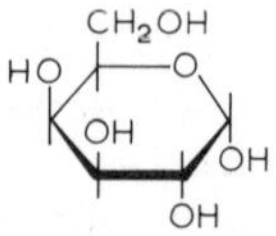

α-D(−) galactopyranose

It is usual to omit the hydrogen atoms when using the Haworth method of presentation.

The Haworth model is a convenient approximation of the actual configuration of carbohydrate molecules. In fact, the pyranose ring is not planar. Conformational formulae showing the approximate shape of the molecule have been proposed.

Fructose (or levulose), the so-called "fruit sugar", is a ketose and is found in honey and fruit juices in its pyranose form; however, when in combination with other sugars, such as in the case of sucrose or in polysaccharides, its cyclic form is furanose.

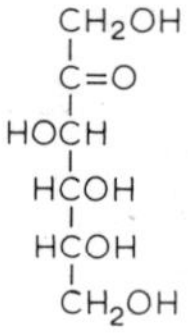

open-chain form

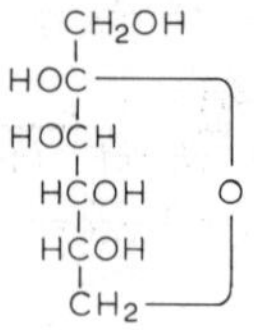

α-D(−)fructopyranose

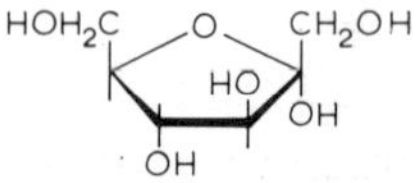

α-D(−)fructofuranose

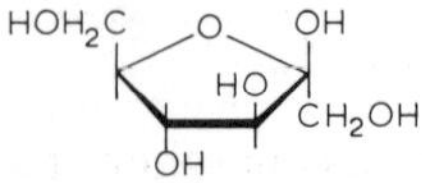

β-D(−)fructofuranose

Pentoses also occur as furanoses or pyranoses. Examples of important pentoses are discussed below.

D-Ribose and its deoxy derivate (see Section 6-4-2) are widespread both in the plant and animal kingdom as important constituents of nucleotides (Section 6-4-4). *L-Arabinose* is found free as well as combined in plant gums and hemicelluloses (Chapter 9). *D-Xylose* or wood sugar is obtained by hydrolysis of structural polysaccharides in wood, corn cobs, hulls, straw, etc.

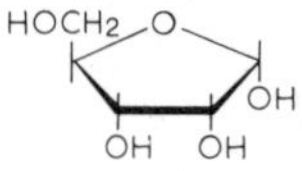

α-D-ribose

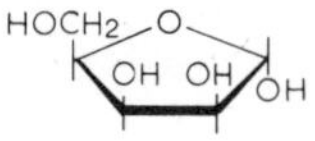

α-L-arabinose

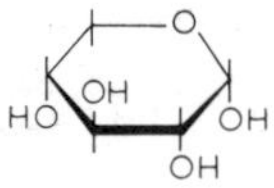

α-D-xylose

Two heptoses occur in the free state, *D-manno-heptulose* (in avocado) and *Sedoheptulose* (in succulent plants). Both are ketoses with the carbonyl group on the second carbon atom. Sedoheptulose is an intermediate product of photosynthesis and of carbohydrate metabolism.

Tetroses and trioses are important intermediates in the biosynthesis and catabolism of sugars and will be discussed in connection with these processes.

6-3. CHARACTERISTIC REACTIONS

6-3-1. *Reactions of the carbonyl group.* (a) *Enolization of sugars — The* Lobry De Bruyn and Alberda Van Ekenstein *transformation.* From the epimeric sugars, those which differ from each other only by their configuration at carbon atom 2 can be transformed into each other by heating them in alkaline solution. Thus, when D-glucose is treated accordingly, a mixture of three sugars of the following composition is obtained: D-glucose, 65%; D-fructose, 31%; D-mannose, 2.4%.

The mechanism of this transformation is believed to involve an intermediate stage of *enediol* formation. Rearrangement of the enediol results in the production of ketose or the epimeric aldose:

```
   HCO                 HC-OH                HCO
    |       <==>        ||       <==>        |
 H-C-OH                 C-OH              OH-C-H
    |                   |                    |
 aldose A             enediol             aldose B
 (glucose)                                (mannose)

                        ||
                       <==>

                      H2C-OH
                        |
                        C=O
                        |
                      ketose
                    (fructose)
```

(b) *Oxidation; aldonic acids.* All monosaccharides which possess a potential free carbonyl group have the ability to reduce *Fehling's solution* and therefore this solution may be used as an excellent qualitative and quantitative reagent for these sugars. When Fehling's solution, which is an alkaline solution of $CuSO_4$ in sodium potassium tartrate buffer, is heated in the presence of such a monosaccharide, the cupric ion is reduced to the cuprous form and appears as the red cuprous oxide.

The mechanism of this reaction is based on the fact discovered by Von Euler that all sugars with a free carbonyl group are easily transformed, by heating them in alkaline solution, into enediols (see above), strongly reactive compounds oxidizable by oxygen and by other oxidizing agents, and therefore termed *reductones.*

$$\underset{\text{OH}}{\underset{|}{\text{RC}}}=\underset{\text{OH}}{\underset{|}{\text{C}}}\cdot\text{CHO}$$

reductone

These reductones have much in common with the structure of ascorbic acid (vitamin C), as will be seen later in this chapter. These compounds, having a strong reducing power, can reduce ions such as Ag^+, Hg^{2+}, Cu^{2+}, Bi^{3+} or $Fe(CN)_6^{3-}$ and by means of this the sugars are oxidized into complicated sugar acids. There are several modifications of Fehling's method, such as the Benedict, Somogyi, Hagedorn-Jensen, or the Willstaetter-Schudel methods, but all of them are based on the same principle.

Through oxidation of C_1 carbonyl group aldoses are transformed into the corresponding *aldonic* acids:

$$\underset{|}{\text{H}-\text{C}=\text{O}} \xrightarrow{[\text{O}]} \underset{|}{\text{HO}-\text{C}=\text{O}}$$

aldose (glucose) aldonic (gluconic) acid

The oxidation of glucose to gluconic acid is of some importance in biotechnology. This reaction is readily carried out in the presence of *glucose oxidase*, an enzyme widely distributed in nature. Commercial preparations of this enzyme, known also as "*notatin*", are obtained from fungi. Although commonly known as oxidase, it is really a dehydrogenase requiring flavin–adenine dinucleotide (FAD) as a coenzyme. The hydrogen removed from the sugar is finally transferred to molecular oxygen, with formation of H_2O_2.

In the soft drink industry it is often desirable to remove from the product all traces of dissolved oxygen which could otherwise impair the stability of the beverage. This may be achieved by adding glucose oxidase which uses up the oxygen to oxidize the glucose normally present in such products. Similarly, glucose oxidase may be used to "destroy" the carbonylic group of glucose, when this group may participate in undesirable reactions, such as the Maillard browning reaction. This latter use of glucose oxidase will be described in more detail in connection with non-enzymic browning.

C_1 oxidation of glucose is the basis of the commercial production of *gluconic acid*, a chemical of some use in industry and pharmacy. The reaction is carried out either electrolytically or biologically (aerobic fermentation by Aspergillus niger).

(c) *Reduction, polyols.* Carbonylic substances can be reduced to corresponding alcohols. Consequently, given proper conditions, aldoses and ketoses may be expected to undergo reduction, and yield polyols. Three *hexitols* are widespread among plants:

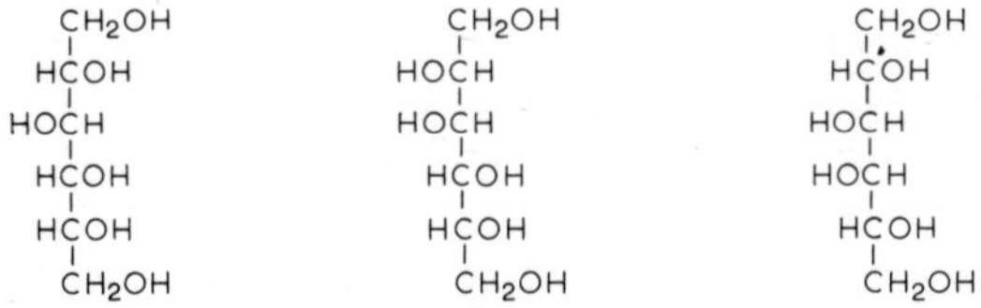

sorbitol	D-mannitol	dulcitol (galactitol)
CH_2OH	CH_2OH	CH_2OH
HCOH	HOCH	HCOH
HOCH	HOCH	HOCH
HCOH	HCOH	HOCH
HCOH	HCOH	HCOH
CH_2OH	CH_2OH	CH_2OH

The structure of *pentitols* may be easily derived from that of the corresponding pentose (Section 6-2). Pentitols are not frequently found in nature. Ribitol is a constituent of riboflavin or Vitamin B_2 (Section 3-3).

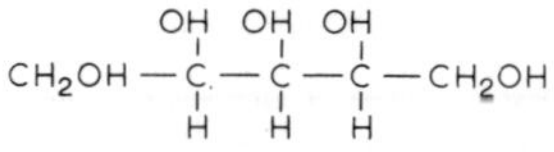

$$CH_2OH-\underset{H}{\overset{OH}{C}}-\underset{H}{\overset{OH}{C}}-\underset{H}{\overset{OH}{C}}-CH_2OH$$

ribitol

Tetritols are also uncommon, but the only triol, *glycerol*, is one of the most important polyols, being an essential constituent of all fats and oils (see Lipids).

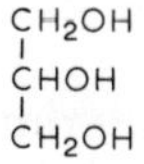

CH_2OH
CHOH
CH_2OH

glycerol

A different class of polyols, the *inositols*, are widely distributed and biologically important. Inositols are hexahydroxy derivates of cyclohexane and while they are not direct products of hexose reduction, they are usually regarded as hexose derivatives in view of the similarity of structure. The most abundant stereoisomer of inositol is *myo-inositol.*

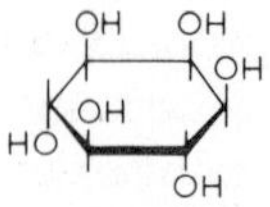

myo-inositol

The hexaphosphoric ester of myo-inositol, known as *phytic acid*, is abundant in the plant kingdom as the magnesium–calcium salt, *phytin*. Phytin is partially saponified to lower phosphates by a special phosphatase, *phytase*. Phytic acid is believed to interfere with the adsorption of calcium and iron from the gastro-intestinal tract, by forming insoluble salts with these cations.

Myo-inositol is also a component of the widespread lipid phosphatidyl inositol (see Lipids). Inositols are crystalline, water soluble and usually sweet.

Polyols are generally sweet. *Sorbitol* is used as a sugar substitute in special foods for diabetics. Its absorption in the intestinal tract is slower than that of glucose and therefore the ingestion of a food containing sorbitol does not result in immediate rise of blood sugar level, although after its absorption, sorbitol is converted to fructose and metabolized exactly in the same way as this sugar.

Fatty acid esters of sorbitol are valuable in food industry as emulsifying agents, whereby the numerous hydroxyl groups of the polyol residue provide the necessary hydrophylic properties while the fatty acid moiety acts as the fat binding group.

Enzymes which catalyse the reduction of sugars into corresponding polyols have been found in nature. These are "aldose reductases" or more correctly, *polyol dehydrogenases*, acting in reverse. Commercially, sorbitol is produced by high pressure hydrogenation of glucose in the presence of nickel as a catalyst.

Reduction to polyols is an important step in the interconversion of sugars in nature. For example, glucose is reduced to sorbitol by a specific dehydrogenase (requiring NADP). Sorbitol can now be oxidized to fructose by another specific dehydrogenase (requiring NAD). This sequence provides a route for the conversion of glucose to fructose and vice versa.

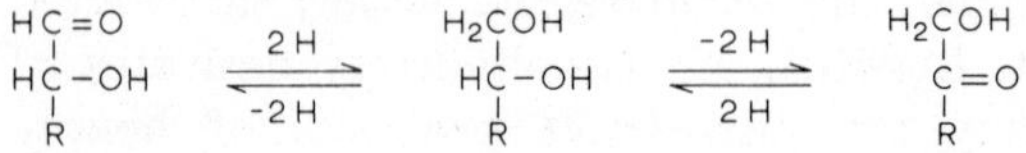

$$\underset{\text{aldose}}{\begin{array}{l} HC{=}O \\ HC{-}OH \\ R \end{array}} \underset{-2H}{\overset{2H}{\rightleftharpoons}} \underset{\text{polyol}}{\begin{array}{l} H_2COH \\ HC{-}OH \\ R \end{array}} \underset{2H}{\overset{-2H}{\rightleftharpoons}} \underset{\text{ketose}}{\begin{array}{l} H_2COH \\ C{=}O \\ R \end{array}}$$

(d) *Glycoside formation.* One of the most important characteristics of the carbonyl group of sugars is their ability to undergo condensation with alcohols, by elimination of one molecule of water. The resulting compound is a glycoside.

$$\text{H(OH)C} \langle \text{O} \rangle + \text{HOR} \rightleftharpoons \text{H(OR)C} \langle \text{O} \rangle + H_2O$$

glycoside

The glycosidic bond is different from a true ether bond in that it is highly susceptible to hydrolysis under relatively mild conditions. When the alcohol moiety is itself a sugar, the glycoside is a disaccharide. Continuation of the condensation process yields polysaccharides. These will be discussed in the next section. When the alcohol is not a sugar, the non-sugar moiety of the glycoside is termed *aglycon*. Many natural substances of considerable significance in food technology are glycosides (bioflavonoids, anthocyanine pigments). A number of natural glycosides have important physiological properties (some plant hormones, saponins). Some are used in pharmacy (e.g. the cardiac glycosides). The physiological properties of these glycosides usually reside in the aglycon while the sugar moiety may often be important for the stabilization of the aglycon and for controlling the adsorption of the glycoside in the body.

6-3-2. *Reactions of the hydroxyl groups.* Among the physical properties associated with the plurality of hydroxyl groups one can cite: solubility of sugars in water, their ability to associate through hydrogen bonds, the water adsorption capacity of polysaccharides and probably the sweetness of sugars. Chemically, the hydroxyl groups may be expected to behave according to their position, i.e. to their being primary, secondary or tertiary hydroxyls.

(a) *Sugar esters.* Acetylation and methylation of carbohydrates have been used as research tools in the elucidation of the structural chemistry of sugars. Other carbohydrate esters (cellulose acetate, surface active esters) are produced industrially. However, the most important esterification reactions in which sugars are involved are undoubtedly the carbohydrate phosphorylation reactions in vivo. Formation, decomposition and transformation of phosphate esters of saccharides are of central importance in carbohydrate metabolism. Harden and Young were the first to observe the presence of a phosphorylated sugar during the catabolic process of fermentation: when inorganic phosphate was added to a fermenting glucose solution the phosphate became quickly bound to the glucose and the ester they isolated was found to be a fructose diphosphate. This phosphoric ester of fructose, as well as the other corresponding esters, have been named after various investigators. Here are the formulae of four such hexose esters:

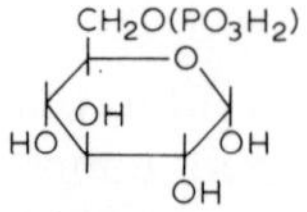

glucopyranose-6-phosphate
(Robison ester)

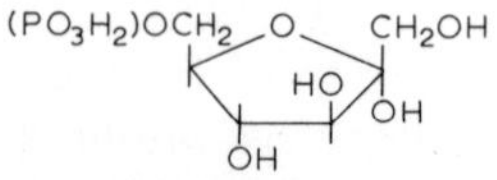

fructofuranose-6-phosphate
(Neuberg ester)

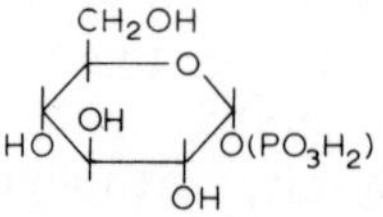

α-glucopyranose-1-phosphate
(Cori ester)

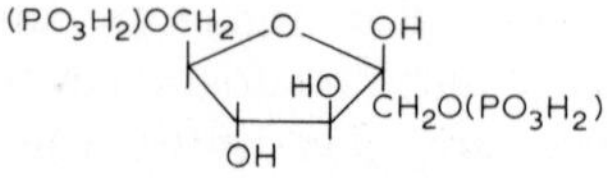

β-fructofuranose-1,6-diphosphate
(Harden-Young ester)

The hexose phosphorylation is catalyzed by the enzyme *hexokinase*, which is found in all higher plants and animals and which has been isolated in its crystalline form from yeast; its behavior has been investigated in particular in potatoes and in the leaves of spinach. In the same way as many other enzymes, hexokinase acts jointly with ATP — adenosine triphosphate. When *hexokinase* reacts with glucose, ATP transfers one of its molecules of phosphoric acid to the sugar, and thus transforms itself into ADP:

$$\text{glucose} + \text{ATP} \xrightarrow{\text{hexokinase}} \text{glucose-6-phosphate} + \text{ADP}$$

The standard free energy of hydrolysis of the phosphate ester bond in glucose-6-phosphate is 3.3 kcal/mol.

Glucose-6-phosphate, in its turn, can be transformed into glucose-1-phosphate by means of another enzyme, *phosphoglucomutase* and by still another, *phosphoglucoisomerase*, into fructose-6-phosphate. The overall transformations of the phosphorylated hexoses can be represented as follows:

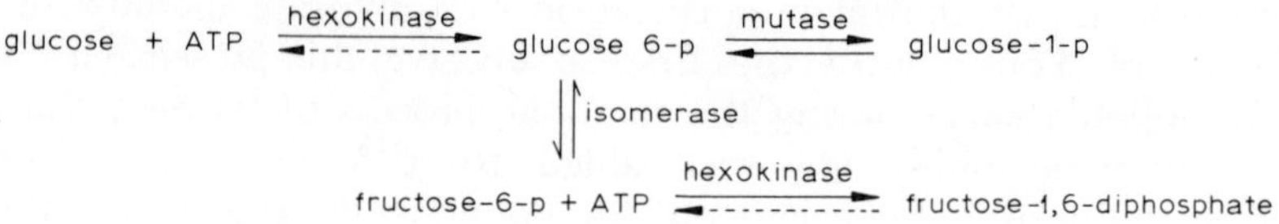

$$\text{glucose} + \text{ATP} \underset{}{\overset{\text{hexokinase}}{\rightleftharpoons}} \text{glucose 6-p} \overset{\text{mutase}}{\rightleftharpoons} \text{glucose-1-p}$$

$$\text{glucose 6-p} \;\overset{\text{isomerase}}{\rightleftharpoons}\; \text{fructose-6-p}$$

$$\text{fructose-6-p} + \text{ATP} \overset{\text{hexokinase}}{\rightleftharpoons} \text{fructose-1,6-diphosphate}$$

In other words, such a transformation involves three enzymes and two molecules of ATP. These are the very first steps of the catabolic processes, respiration and fermentation. In a reverse direction, these transformations take place also in the anabolic processes of photosynthesis.

Enzyme-catalyzed interconversions of other hexose-monophosphates, such as mannose-6-phosphate and galactose-1-phosphate, are also known; however, the enzymes engaged in some of these transformations have not been sufficiently purified except one enzyme (*uridyl transferase*) involving uridine diphosphate glucose (UDPG) as a cofactor, which converts galactose-1-phosphate to glucose-1-phosphate by the Walden inversion.

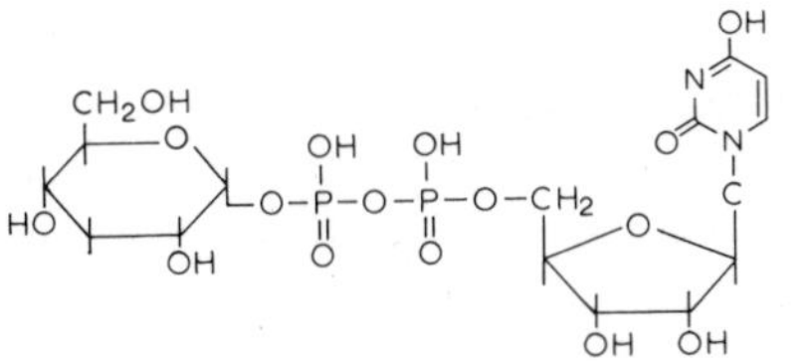

uridine diphosphate glucose (UDPG)

(b) *Oxidation, uronic acids.* Selective oxidation of the primary carbinol group (e.g. the C-6 position of glucose, galactose, etc.) is carried out by enzymic methods as well as with the help of some inorganic catalysts. The products of such oxidation are uronic acids.

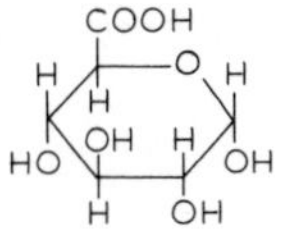

α-glucuronic acid

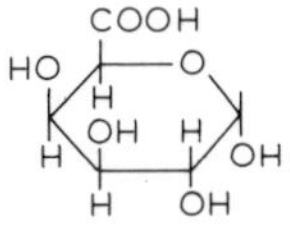

α-galacturonic acid

Uronic acids are widespread in nature, as building stones, of many polysaccharide derivatives. Galacturonic acid is the basic unit of which pectic substances are built (see Chapter 9). Glucuronic acid is an important detoxifying agent, which combines with a number of toxic substances through glycosidic bonds. The complexes are eliminated in the urine.

Oxidation of the secondary hydroxyls yields ketones. This reaction is of some interest in chemical synthesis.

6-4. OTHER IMPORTANT DERIVATIVES

6-4-1. *Ascorbic acid.* As far back as in 1747, a doctor in the British Navy, Captain Lind, showed that the dangerous disease called *scurvy*, which frequently prevailed with seamen who were at sea for many months, was caused only because of the lack of fresh fruit and vegetables in their diet. Lind also found, through long empirical tests, that the best way to prevent scurvy was to add to the daily diet of men at sea some lemon or lime juice. Since 1804, the British Navy had strict instructions never to sail without lime juice — the British sailors have since been nicknamed the "limies". Attempts to isolate and characterize the antiscorbutic factor in lemon juice began more than a hundred years later. Zilva obtained highly concentrated active preparations from lemons. He found that the antiscorbutic factor was a water soluble, strongly reducing substance, closely related to the sugars. Szent-Györgyi succeeded in crystallizing the compound, to which he assigned the name "*hexuronic acid*". Later the substance was named *L-ascorbic acid* and its structure was shown to be as follows:

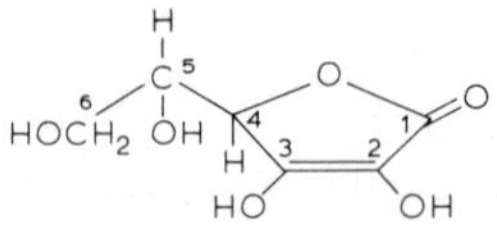

L-ascorbic acid

L-ascorbic acid is a vitamin (see Section 3-3), *Vitamin C.* Man and only a few other mammals (monkeys, guinea pigs) require vitamin C, since they lack the enzyme necessary to perform one of the steps in its biosynthesis.

Vitamin C is widespread in the plant kingdom, but some fruits are exceptionally rich in it. One of the richest fruit sources of ascorbic acid is the *acerola* (West Indian) *cherry*, which contains up to 2000 mg/100 g. Other important sources are: rose hips, paprika, pine needles, guava (300 mg/100 g), black currant (210), parsley (190), broccoli (120), green pepper (120), citrus fruit (35-50), potatoes (30).

L-ascorbic acid crystallizes in white, odorless plates, somewhat acid in taste, melting at 190-192° C, is exceedingly soluble in water (1 g in 3 ml) but insoluble in most organic solvents.

Ascorbic acid is optically active $[\alpha]_D^{20} = +23°$. The vitamin C activity of L-ascorbic acid and of its oxidized form, *dehydroascorbic acid*, is the same. The D isomer, *isoascorbic acid*, is only 1/5 to 1/20 as potent.

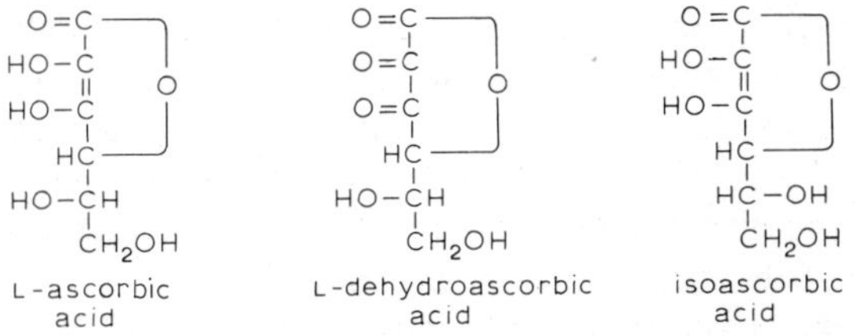

```
O=C ──┐          O=C ──┐          O=C ──┐
HO-C   O         O=C    O         HO-C   O
HO-C             O=C              HO-C
HC ───┘          HC ───┘          HC ───┘
HO-CH            HO-CH            HC-OH
CH2OH            CH2OH            CH2OH
```

L-ascorbic acid   L-dehydroascorbic acid   isoascorbic acid

Ascorbic acid behaves as a monobasic acid. It reacts with sodium bicarbonate to form sodium ascorbate $C_6H_7O_6Na$ with the liberation of CO_2. The acidic reaction is due to the enediol configuration.

The most important characteristic of ascorbic acid is its ability to readily undergo reversible oxidation to dehydroascorbic acid. The physiological activity of vitamin C depends largely upon this reversible oxidation. Oxidation of ascorbic acid takes place in the presence of molecular oxygen and is greatly accelerated by traces of metals, especially copper. This oxidation is also catalyzed by a number of non-specific oxidases as well as the specific *ascorbic acid oxidase (ascorbinase)* containing copper. The end products of the reaction are dehydroascorbic acid and H_2O or H_2O_2 (depending on the catalyst).

Dehydroascorbic acid can be easily reduced to ascorbic acid by H_2S. Biologically, this reduction is catalyzed by *dehydroascorbic acid reductase*, in the presence of *glutathione* (see Sections 2-5-4 and 2-5-5). The sulfhydryl groups of glutathione act as hydrogen donors with the formation of disulfide bonds.

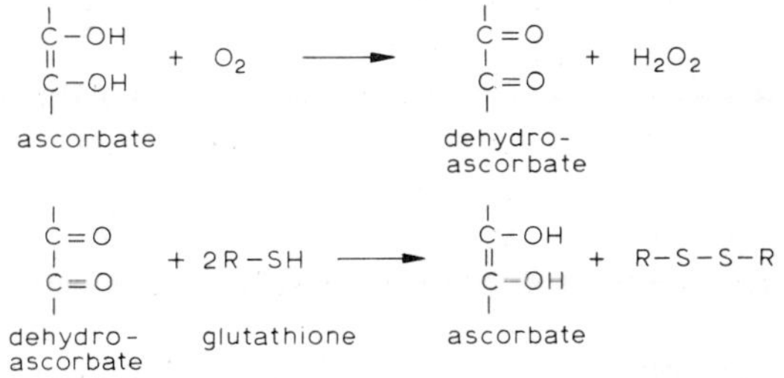

$$\text{C-OH}=\text{C-OH (ascorbate)} + O_2 \longrightarrow \text{C=O}-\text{C=O (dehydroascorbate)} + H_2O_2$$

$$\text{C=O}-\text{C=O (dehydroascorbate)} + 2\,\text{R-SH (glutathione)} \longrightarrow \text{C-OH}=\text{C-OH (ascorbate)} + \text{R-S-S-R}$$

The participation of ascorbic acid and dehydroascorbic acid in biological electron transport (oxidation–reduction reactions) seems to be associated mainly with hydroxylation processes. Ascorbic acid is essential in the *hydroxylation of proline to hydroxyproline.* It was mentioned earlier (Section 2-3) that the relatively uncommon hydroxyproline participates in the composition of the connective tissue protein *collagen.* Many of the lesions typical to scurvy are due to abnormalities in the connective tissue. Experimental evidence of the participation of ascorbic acid in other biological hydroxylation processes (hydroxylation of the aromatic ring in the metabolism of tyrosine, biogenesis of steroids, etc.) has been presented.

Reactions leading to the destruction of vitamin C during food processing have been studied extensively. Ascorbic acid is relatively thermostable. However, dehydroascorbic acid is rapidly destroyed by heat. In neutral media the lactone ring of the dehydroascorbic acid is entirely destroyed within 10 minutes at 60° C. This breakdown is not an oxidative phenomenon and has been shown to take place under anaerobic conditions. The product of this transformation is the dienolic form: 2,3-diketogulonic acid.

O=C–O(ring)–; O=C; O=C; HC–(ring); HO–CH; CH_2OH → COOH; O=C; HO–C; ‖ C–OH; HO–CH; CH_2OH

dehydroascorbic acid → 2,3-diketogulonic acid

Extensive changes, especially in color and flavor, occurring in fruit and vegetable products during storage, run parallel with the progressive decrease in the amount of ascorbic acid. Darkening of citrus juices, for instance, during storage, has been shown to occur after ascorbic acid has been totally and irreversibly oxidized. In fact, one of the forms of browning of food products, as will be shown later (Chapter 10), involves the transformation of ascorbic acid and its oxidation products into furfural by dehydration and decarboxylation. The furfural so formed polymerizes and gives rise to dark colored products or it can enter into further combinations with amino acids, etc.

To prevent oxidation of vitamin C when dealing with foods containing it, several precautions must be strictly observed. In the first place, the enzyme ascorbinase should be inactivated; this is especially important when dehydrating fruits and vegetables, and it can be achieved by properly blanching such foods. Steam blanching is preferable to scalding in hot water since vitamin C is lost to a great extent due to leaching in large quantities of water. Secondly, oxygen should be excluded as far as possible during the preparation and processing of food products: whenever possible, deaeration or exhausting of the food products should be carried out. Food products which are essential for their vitamin C content should be canned, if possible, in plain tin cans, where the reducing conditions help to prevent the autoxidation of ascorbic acid. In some cases, antioxidants can be used to prevent oxidation. It has been reported, as a result of a series of oxidation–reduction potential studies, that D-isoascorbic acid, the antiscorbutic activity of which is only one-twentieth that of L-ascorbic acid, is a strong antioxidant, for it oxidizes more rapidly than vitamin C in food products, thus protecting the latter from deterioration.

Taking all the above factors into consideration (absence of air, proper inactivation of ascorbic acid oxidase, freedom from traces of copper in processing equipment) manufacturing conditions can be worked out under which the losses of vitamin C will be reduced to a mere minimum.

Ascorbic acid and its D-isomer (isoascorbic acid) are used extensively as antioxidants in foods. One of the most widely practised applications is in the prevention of enzymic oxidative browning of some foods (apples, potatoes, artichoke, eggplant, avocado).

Several methods exist for preparing vitamin C synthetically. One technically possible and economically worthwhile method, now largely used in industry and yielding some 15% of the original starting material, is that proposed by Reichstein et al. Although, theoretically, the synthesis can begin with glucose, technically the process starts with L-sorbose, according to the following scheme:

D-glucose		sorbitol		L-sorbose (110 g)	
CH_2OH		CH_2OH		CH_2OH	
HOCH		HOCH		C=O	
HOCH	$\xrightarrow{H_2}$	HOCH	$\xrightarrow{\text{oxidation by sorbose bacterium}}$	HOCH	$\xrightarrow{\text{acetone}}$
HCOH		HCOH		HCOH	
HOCH		HOCH		HOCH	
CHO		CH_2OH		CH_2OH	

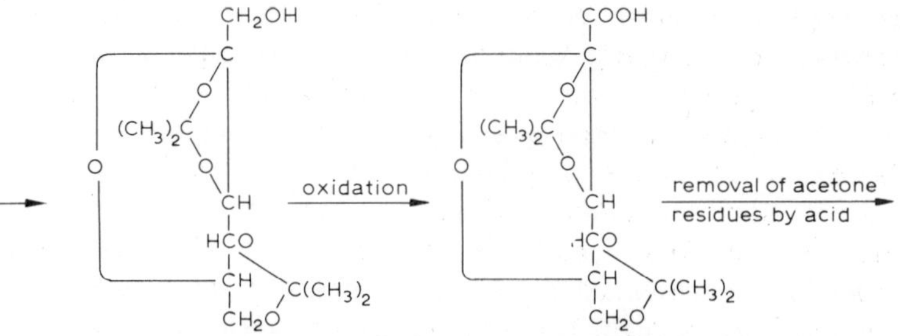

diacetone-L-sorbose (100 g) → oxidation → diacetone-2-ketogulonic acid (85 g) → removal of acetone residues by acid →

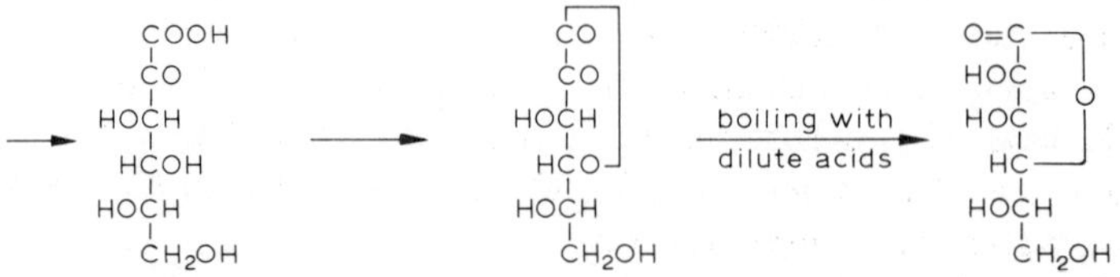

L-xylo-2-ketohexonic acid (46 g) → lactone of xylo-2-ketohexonic acid → boiling with dilute acids → L-ascorbic acid (24 g)

The estimation of vitamin C activity was, for a long time, performed by bioassays with guinea pigs. Tillmans found a specific chemical method for the titration of ascorbic acid using *2,6-dichlorophenol indophenol*, a blue substance which is specifically oxidized into its colorless leuco form.

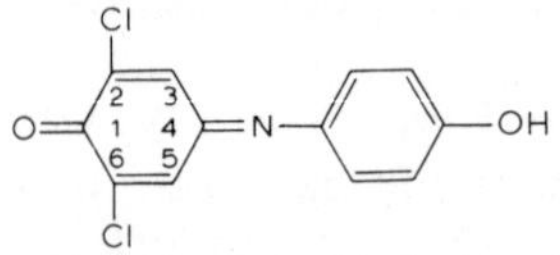

2,6-dichlorophenol indophenol

Notwithstanding the fact that ascorbic acid has been the most studied vitamin, there are still many unsolved problems regarding its biogenesis, its functions in plants and in animals, the great variations of its content in different plants, as well as the diversity in the capability of animals to synthesize their own vitamin C. Even mechanisms connected with the activity of ascorbic acid are largely unsolved.

Furthermore, it is still unknown why one and the same variety of a given plant contains varied amounts of ascorbic acid, and even in different parts of one single fruit its content varies. It has been established, for instance, that in many fruits there exists a definite gradient of ascorbic acid which declines from the epicarp towards the endocarp: several varieties of Italian oranges tested showed the following contents in mg per 100 g:

in the flavedo (yellow peelings)	175-292 mg
in the albedo (white peel)	86-194 mg
and in fruit flesh	44.9-73.2 mg

A similar distribution of vitamin C has been found in the lime (Citrus limetta Risso), while the guava grown in Israel showed the following gradient relationship between the epicarp, the flesh and the center of the fruit — 9 : 4 : 1.

It has been shown that in bruised or cut portions of fruits or vegetables, the amount of vitamin C is greatly augmented after a few days. In sliced potatoes, for instance, the amount of vitamin C rises from 10.8 mg to 24.1 mg per 100 mg in three days. This traumatic formation of ascorbic acid is considered to be a "defense mechanism".

While most other vitamins are required by man for his well being in γ quantities, the recommended daily requirement for ascorbic acid amounts to 60 mg. A daily intake of 10 mg is effective in preventing scurvy. Unlike many other vitamins, vitamin C cannot be stored in the body. Any excess in intake is eliminated in the urine. Recently, it has been claimed that ingestion of much larger quantities, a few grams per day, results in positive prevention of the *common cold*. This view, supported by the famous chemist Pauling (holder of two Nobel Prizes), has come under severe criticism. Obviously, such large quantities cannot be secured from normal dietetic sources and must be supplied as synthetic L-ascorbic acid.

6-4-2. *Deoxy sugars.* The most important deoxy derivative of monosaccharides is *2-deoxy-D-ribose*, found in the composition of deoxyribonucleic acid (DNA).

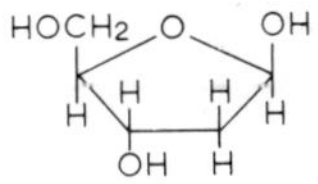

$HOCH_2$ O OH
H H H H
OH H

2-deoxy-D-ribose

6-4-3. *Amino sugars.* A number of biologically important polysaccharides (Chapter 8) contain nitrogen. On hydrolysis, they yield *amino sugars*, two of which are relatively abundant: *D-glucosamine* and *D-galactosamine*; in both compounds, an amine group replaces the hydroxyl on carbon atom 2 of the corresponding monosaccharide.

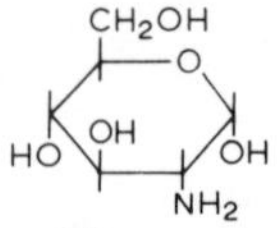

α-D-glucosamine (chitosamine) α-D-galactosamine (chondrosamine)

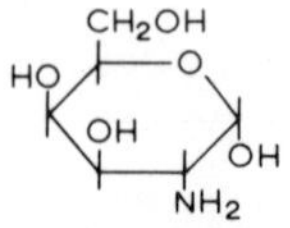

6-4-4. *Glycosylamines, nucleosides, nucleotides.* Aldehydes react with compounds containing an amino group, according to the following well known reaction:

$$R{-}CH{=}O + H_2NR \longrightarrow [R{-}CH(OH){-}NHR] \longrightarrow R{-}CH{=}NR + H_2O$$

Schiff base

Sugars form similar addition compounds with amines, amino acids, etc. As will be shown later, this process is the first step of Maillard's reaction, which leads to the most common type of browning discoloration of foods. When applied to the ring form of the sugar, the addition of amino compounds may be represented as follows:

sugar + HN–R ⟶ glycosylamine (HNR) + H_2O

The similarity between this reaction and the formation of glycosides (Section 6-3-1) is evident.

A group of natural substances of vital importance, the *nucleosides* are in fact glycosylamines where the sugar is *ribose* or *deoxyribose*, and the nitrogenous partner is one of the heterocyclic bases known as *purines* and *pyrimidines*. Two purines (*adenine* and *guanine*) and three pyrimidines (cytosine, uracil, thymine) are the major bases in nucleosides.

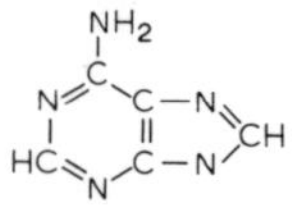

adenine 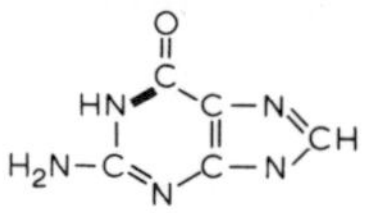 guanine

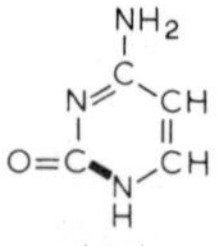

cytosine 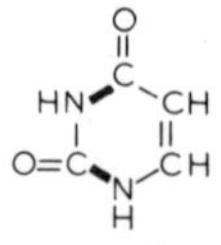 uracil 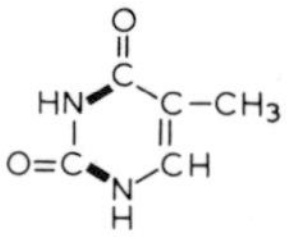 thymine

The bond between the base and the sugar is illustrated for the case of the nucleoside *adenosine* (adenine–ribose).

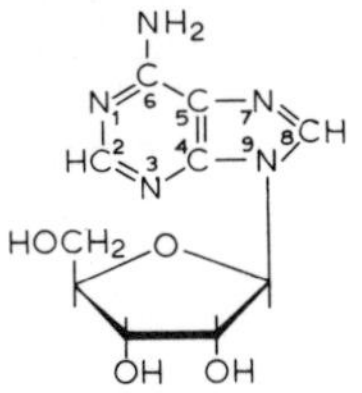

adenosine
(9-β-D-ribofuranosyladenine)

If one of the free hydroxyl groups of the pentose is esterified to phosphoric acid, the resulting compound is a *nucleotide.* The role of several nucleotides as coenzymes (NAD, FAD, etc.), or biological agents of energy transfer (ADP, ATP) has been discussed in the previous chapters. *Nucleic acids* are highly organized macromolecules consisting of linear polymers of different nucleotides. Nucleic acids are of the utmost importance in the vital processes of protein biosynthesis and transfer of genetic information.

In this connection, it is appropriate to mention the so-called *flavor nucleotides.* Two mononucleotides, *5′-inosine monophosphate* (IMP) and *5′-guanosine monophosphate* (GMP) are potent enhancers of meaty flavors in foods. They occur naturally in fish, shellfish and mushrooms and they are produced commercially by enzymic hydrolysis of yeast RNA. They are used extensively in dry soup mixtures, gravies, processed meat, fish and seafood, simulated meat products, etc., usually in combination with protein hydrolysates or with another flavor enhancer, *monosodium glutamate* (MSG).

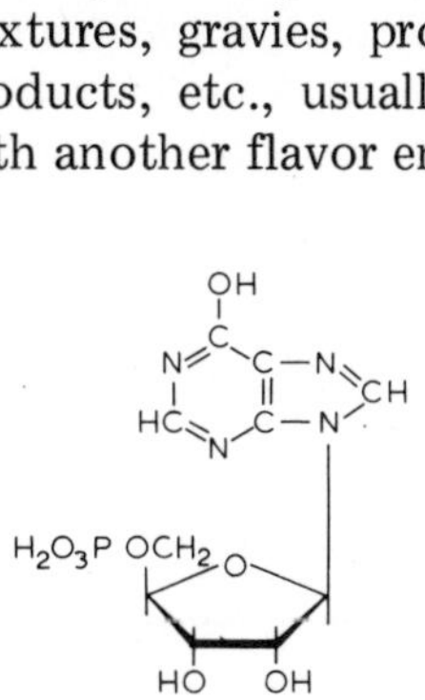

5′-inosine monophosphate (IMP)

SELECTED BIBLIOGRAPHY TO CHAPTER 6

Pigman, W. (Ed.), The Carbohydrates, Chemistry, Biochemistry, Physiology, Academic Press, New York, 1958.

Sebrell, W.H. Jr. and Harris, R.S. (Eds.), The Vitamins, 2nd edn., Vol. 1, Academic Press, New York, 1967.

Staněk, J., Cerny, M., Kocourek, J. and Pacák, J., The Monosaccharides, Academic Press, New York, 1963.

CHAPTER 7

CARBOHYDRATES — OLIGOSACCHARIDES

When two or three molecules of monosaccharides (monoses) are joined together they form disaccharides or trisaccharides, respectively. Such combined sugars containing from two to ten monose residues are given the general name of *oligosaccharides* (oligo = few). These can be formed of either the same or of different monoses. Thus, for instance, maltose is formed of two glucose molecules, while in sucrose one molecule of glucose is bound to one molecule of fructose.

7-1. DISACCHARIDES

Disaccharides may be regarded as glycosides in which one molecule plays the role of an aglycone bound to another monose molecule with the elimination of H_2O. In accordance with this definition, when studying the structure of disaccharides, one has to examine in each case the following points:

(1) Which monoses constitute a given disaccharide? This can usually be found by acid hydrolysis and subsequent determination of the resulting constituents: if only one monose is found the indication is that two molecules of the same monose created the disaccharide (as in the case of maltose mentioned above). In the case of sucrose, however, equal quantities of glucose and fructose will be found in the resulting mixture after hydrolysis.

(2) In what configuration do they appear, α or β? When one of the moieties is glucose, this can be determined, according to Armstrong, by using the enzymes, α- and β-glucosidases. If the bond between the two molecules of the disaccharide is an α-bond it will be hydrolyzed by *α-glucosidase* (maltase), which is obtained from germinating barley. If the bond has a β-configuration, the hydrolysis will be catalyzed by *β-glucosidase* (emulsin) found in the seeds of Rosaceae.

(3) Which monose plays the role of the aglycone?

(4) At which carbon atom has the bond between the two monoses been formed? Is it a 1 : 1 bond or 1 : 4, or 1 : 6 bond?

(5) Finally, what is the type of the rings in both monoses; are they pyranose or furanose? The latter points are usually elucidated by methylation or acetylation procedures applied to the disaccharide in question followed by hydrolysis to single monoses.

There are only three disaccharides found in nature in the free state: sucrose in sugar and sugar beet and in many fruits, lactose in milk, and trehalose which has been identified as the principal carbohydrate component of the blood of some insects. All other disaccharides are "building stones" in many polysaccharides, such as maltose, the product of starch hydrolysis, cellobiose — from cellulose, etc.

The structure of the disaccharides is presented here, some examples of the various types being given:

(a) The case when the two monoses are the same, e.g. *maltose*:

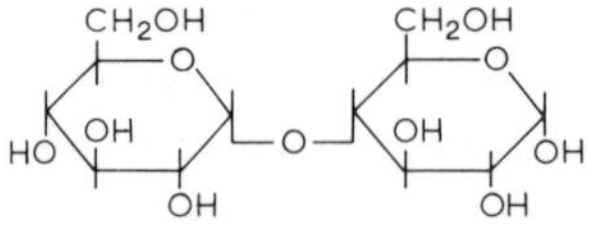

maltose: 4-D-α-glucose-α-D-glucopyranoside

Maltose is obtained when starch is hydrolyzed either by acid or by the enzymes maltase or diastase, contained in germinating barley. This disaccharide is built of two molecules of α-glucose, bound in the position 1 : 4, i.e. carbon atom 1 of one glucose molecule is bound in a glycosidic bond to carbon atom 4 of the second molecule. It is evident that the carbonyl carbon atom of the second molecule is free, hence the maltose itself can have two stereoisomeric configurations, α and β. Indeed, maltose can undergo mutarotation (from $[\alpha]_D + 130°$ to $[\alpha]_D + 112°$) and is found to be a reducing sugar, i.e. it reduces Fehling's reagent.

(b) *Cellobiose* is also built of two glucose molecules. However, the "gluconic" one is β-glucose and, therefore, the glycosidic bond, although 1 : 4, is a β-bond:

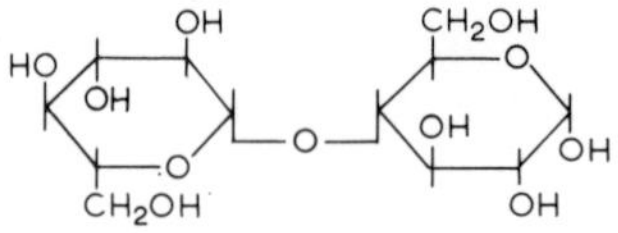

cellobiose: 4-D-glucose-β-D-glucopyranoside

Cellobiose is the building stone for cellulose, reduces Fehling's solution, and can exist in both α and β configurations. It does not exist in nature in the free state.

(c) The case of two identical monoses, both glucose, with a β-glycosidic bond in a different position exists in the disaccharide gentiobiose.

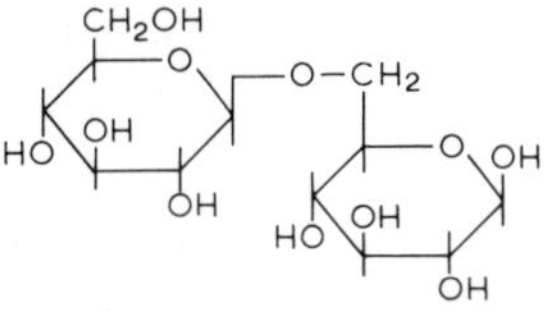

gentiobiose: 6-β-D-glucose-D-glucopyranoside (β form)

This disaccharide is not found in nature in the free state but is a part of the cyanogenic glucoside amygdalin in bitter almonds. It is also a reducing sugar and the bond between the two monoses is 1 : 6.

(d) To end this series of disaccharides built of glucose alone is the example of trehalose, a non-reducing sugar since its glucosidic bond is 1 : 1. Both glucoses are of the α-configuration. Trehalose is produced by a number of fungi and yeasts and is found in the blood of insects.

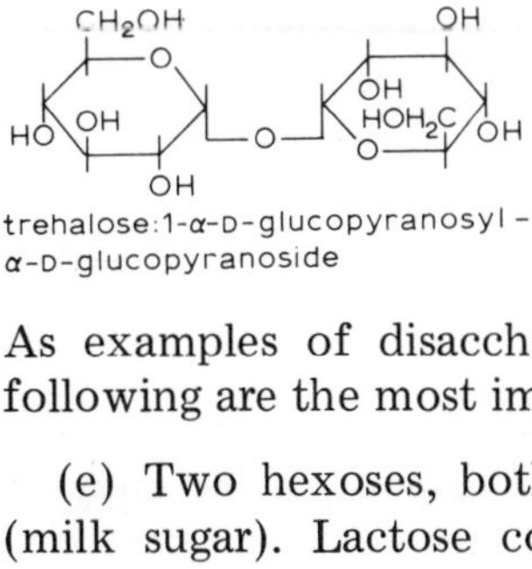

trehalose: 1-α-D-glucopyranosyl-α-D-glucopyranoside

As examples of disaccharides built of two different monoses the following are the most important:

(e) Two hexoses, both pyranoses, are the components of *lactose* (milk sugar). Lactose consists of glucose and galactose. It is the principal sugar in cow's milk, reduces Fehling's reagent and is hydrolyzed by emulsin and by the enzyme *lactase*, which is a β-glucosidase; its structural formula is:

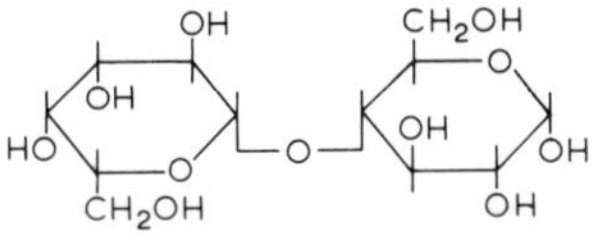

lactose 4-D-glucose-β-D-galactopyranoside

Lactose could, therefore, exist in two configurations, α and β. On oxidation of lactose and subsequent hydrolysis, one obtains galactose and glucuronic acid. This indicates that the free carbonyl group is attached to the glucose and not to the galactose.

(f) The most important disaccharide, widely distributed in nature, is *sucrose.* In its free state it is found in all photosynthetic plants and constitutes the principal disaccharide in the diet of animals. The current world production of sucrose is approximately 80 million tons per year, one-third of which is extracted from sugar beets and two-thirds from sugar cane.

Sucrose is built from two different monoses, α-glucose and β-fructose, the first a pyranose and the second in its furanose form. The glycosidic bond is 1 : 2 between the two carbonyl carbon atoms. Hence, sucrose has no free carbonyl group and therefore does not reduce Fehling's solution and does not combine with sulfur dioxide or with amino acids. This fact is of importance in the relation of sucrose to browning (Chapter 10).

Despite the considerable importance of this disaccharide the structural formula of sucrose was established by Haworth only in 1927, and its first enzymic synthesis performed by Hassid in 1944.

sucrose: 2-D-glucopyranosyl-β-D-fructofuranoside

It is evident that sucrose cannot undergo mutarotation — this is probably the reason why sucrose is easily crystallizable in contrast to many other sugars.

However, sucrose is apt to hydrolyze easily into its components, glucose and fructose, with the aid of weak acids or of the enzyme *invertase.* This phenomenon is called *inversion* and the resulting mixture of the two components — *invert sugar*, due to the fact that the positive sign of the optical rotation of sucrose has been now inverted into the negative sign of the optical rotation of the mixture:

$$\underset{[\alpha]_D = +66.5^\circ}{\text{sucrose}} + H_2O \xrightarrow{\text{inversion}} \underset{[\alpha]_D = +52.5^\circ}{\text{D(+)-glucose}} + \underset{[\alpha]_D = -92^\circ}{\text{D(-)-fructose}}$$

mixture: $[\alpha]_D = -20^\circ$

Honey consists mainly of invert sugar.

7-2. TRISACCHARIDES

Several trisaccharides are found in nature in the free state. *Raffinose* is found in the juice of sugar beet and in cottonseed hulls. It occurs in trace quantities in crystallized commercial beet sugar. Raffinose, on hydrolysis, gives equimolar quantities of D-glucose, D-fructose and D-galactose. Enzymic hydrolysis, with the aid of emulsin, results in sucrose and galactose; in weak acid and with the aid of the enzyme raffinase, melibiose and fructose are formed. From these reactions, the structure of raffinose appears to be as follows:

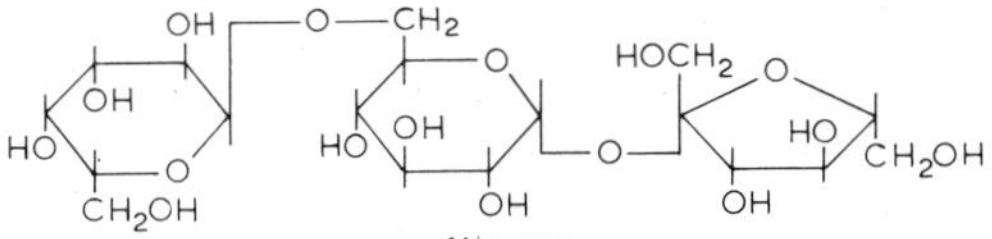

raffinose

Other trisaccharides which occur in nature in the free state include gentianose (in gentian rhizoms), melezitose (in the sweet exudations of several plants, in honey) and plantose (in plants of the Plantago genus).

7-3. TETRASACCHARIDES

One tetrasaccharide, *stachyose*, is widely distributed in the plant kingdom. Stachyose is a hetero-oligosaccharide. Its monomer constituents are D-fructose, D-glucose, and two residues of D-galactose. The fructose, glucose and one of the galactose residues are bound as in raffinose; the additional galactose is connected to this raffinose unit at its galactose terminal by a $(1 \rightarrow 6)-\alpha$ bond.

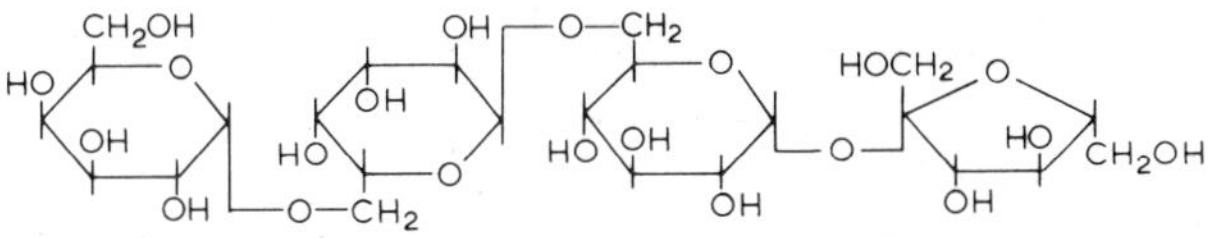

stachyose

Stachyose is one of the main carbohydrates of soybean. Because of its poor digestibility, it is thought to be the major cause of *flatulence* and other gastro-intestinal disturbances experienced as a result of eating soybeans.

7-4. SUGARS IN FOOD TECHNOLOGY

Commercial sucrose (cane sugar, beet sugar), is the principal sugar of the food industry. The culture of sugar cane and exports of sugar are factors of utmost importance in the economy of many countries; dextrose and corn syrups (see Chapter 8) are used in smaller quantity. Lactose is used in specific cases.

Sugars are used in foods, in the first place as *sweeteners*. Of all the sugars contained in foods, D-fructose is known to be the sweetest. It is usual to compare the degree of sweetness of different sugars to sucrose, to which the number 100 has been assigned; fructose is then 173.3 and glucose only 74.3. It is obvious, therefore, that invert sugar is sweeter than the original sucrose from which it is formed: $(173.3 + 74.3) : 2 = 123.8$ for invert sugar. In industry, a number of sucrose syrups are available, partially or totally inverted, having therefore different degrees of sweetness (Table 2). These syrups are

TABLE 2

Degree of Sweetness of Various Sweeteners

Sweetener	Degree of sweetness
Sucrose	100
Fructose	173.3
Glucose	74.3
Lactose	16
Maltose	32
Galactose	32
Saccharin	30,000–50,000
Sodium cyclamate	10,000
Neohesperidin dihydrochalcone	1,000,000

sold to the industry under the name of "liquid sugar", and can be prepared with a high concentration of solids, since fructose has a very high solubility and glucose does not readily crystallize. Artificial sweeteners are used as sugar replacers for dietetic purposes (low-calorie foods, foods for diabetics) or for economical reasons. *Saccharin* (anhydro-O-sulfaminobenzoic acid) is one of the oldest artificial sweeteners. Due to the low solubility of saccharin in the acid form, it is mainly sold and used as the sodium (and sometimes calcium) salt. The bitter after taste of saccharin, to which some people seem to be particularly sensitive, led to the extensive use of another artificial sweetener, *sodium cyclamate.* After several years of recognition as a safe additive, cyclamates were found to be harmful under certain conditions and their use in food was prohibited. The ban on cyclamates encouraged many researchers to look for new non-caloric sweeteners. A long list of such sweeteners, including *flavonoid dihydrochalcones* (Chapter 16) and certain *dipeptides* are now under investigation.

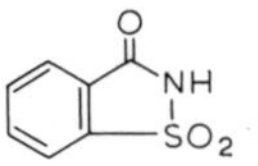

O NH SO_2 $NHSO_3Na$

saccharin 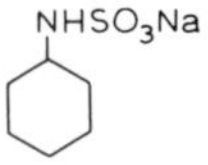 sodium cyclamate

Sugars affect not only the taste but also the appearance and texture of foods. The contribution of sugars to viscosity is important in the consistency, body and mouth-feel of many foods. The high refractive index of concentrated sugar solutions is the reason for the "brilliant" appearance of syrups, jellies, jams and dried fruits. The texture of confectionery products is often based on the presence of solid sugar phases in crystalline or amorphous state.

As a result of their high affinity for water, sugars are efficient water-activity depressing agents. The preservative effect of sugars in jams and candied fruit is based on this property. The water-binding capacity of sugars is also important in the solidification of the pectin-sugar-acid system in jams (see Chapter 9).

With the extremely numerous types of products, the sweets industry has strict requirements for the sweeteners they use. The following is a list of some of their requirements:

(1) The relative sweetness of the various sugars.
(2) Degree of solubility and crystallization.
(3) Specific gravity of the syrups.
(4) Water contents in dry sugars.
(5) Hygrosopicity.
(6) Specific flavor.
(7) Preservation qualities and proneness toward fermenting.
(8) Molecular weight.
(9) Osmotic pressure.
(10) Point of congelation.
(11) Tendency towards browning.

Much depends on the suitable choice of the sugar, taking into consideration its physical properties, in the final product. For instance, if using sucrose, it is very important to know the possibilities of its inversion in the final product. The quality of the various syrups, prepared by the hydrolysis of starch and other natural polysaccharides, depends very much on the structure of these polysaccharides, i.e. whether they are linear or branched, on the size of the molecules, etc. All these qualities open up new possibilities for the manufacturer and modern sweetening materials are indeed tailor-made to the special instructions of the industry.

Sucrose is the most soluble of all sugars and it is quite easy to produce supersaturated solutions, but sucrose is also very easily crystallizable. However, by using invert sugar and various starch syrups, or if the sucrose can be quickly inverted in the product such crystallization, if undesirable, can be prevented. Corn syrups are even better for preventing crystallization, since they do not bring in the factor of *hygroscopicity* which is a fault of invert sugar.

In fruit juices, which are all more or less acid, sucrose undergoes inversion depending on the pH of the juice, the time, and the storage temperature.

In meat products, sugar is used for its ability to lessen the sharpness of salt, and it also prevents discoloration. It is probable that this is due largely to the microorganisms or some enzymatic systems in the meat which are activated by the presence of sugar.

One of the causes of browning (Chapter 10) is the so-called Maillard reaction, which consists of the combination of sugars with a free carbonyl group and amino acids. As a result, some very complicated, brown-colored products, melanoidins, are formed. Hence, by the use of sugars in which this free carbonyl group is absent, browning can be avoided. Sucrose, for instance, will not cause browning in such cases, unless it undergoes inversion. Pure fructose is also inactive in the browning reaction. Glucose, on the other hand, is very active.

SELECTED BIBLIOGRAPHY TO CHAPTER 7

Bailey, R.W., Oligosaccharides, Macmillan (Pergamon), New York, 1965.

Inglett, G.E. (Ed.), Symposium: Sweeteners, The Sci. Publ. Co. Inc., Westport Conn., 1974.

Pasur, J.H., Oligosaccharides, in The Carbohydrates, Chemistry and Biochemistry, W. Pigman and D. Horton (Eds.), Vol. II, pp. 69-137, Academic Press, New York, 1970.
Stanek, J., Cerny, M. and Pacak, J., The Oligosaccharides, Academic Press, New York, 1965.

CHAPTER 8

POLYSACCHARIDES

8-1. DEFINITIONS AND CLASSIFICATION

Polysaccharides are polymers of the simple monoses, joined by glycosidic bonds. By definition, any carbohydrate polymer containing more than ten monosaccharide residues is considered a *polysaccharide.* Most polysaccharides of nature, however, contain close to a hundred monomer residues, and some attain even higher degrees of polymerization, in the range of thousands of residues. Unlike the sugars, polysaccharides are not sweet. They are often responsible, however, for another important sensory characteristic of foods: texture. In many foods, characteristics such as viscosity, mouth feel, consistency, gelation, toughness, fibrousness, etc., are due to polysaccharides. Nutritionally, starch is the most important polysaccharide, since it provides the main source of calories in the human diet.

The polysaccharides may occur as essentially linear long chains (amylose, cellulose), or they may be branched (amylopectin, glycogen). The number of different monosaccharides which participate in the composition of natural polysaccharides is not large. These are some hexoses (mainly D-glucose, D-fructose, D-mannose, D-galactose) and some pentoses (mainly derivatives). The monomers are mainly aldoses and the hydroxyl group on C-1 always participates in the glycosidic bond.

Polysaccharides can be hydrolyzed to shorter chains, oligo- and disaccharides, and finally to the monomer units, by introducing water at the position of the glycosidic bond. Hydrolysis is catalyzed by various enzymes or by acid or alkali. The susceptibility of different polysaccharides to hydrolysis is widely variable.

Those polysaccharides which give, on hydrolysis, only one monomer, are called homopolysaccharides or *homoglycans.* These comprise the most abundant polysaccharides in nature, such as cellulose, starch, galactans, mannans, arabans, xylans, levans and dextrans. Polysaccharides which are resolved by hydrolysis into two or more monomers are named heteropolysaccharides, or *hetero-glycans.* These comprise pectins, hemicelluloses, many mucilages and plant resins.

Polysaccharides are sometimes classified according to their known biological functions, into two main groups:

(a) *"Structural" or "skeletal" polysaccharides* owe their functional properties to the physical characteristics of their macromolecular structure. These characteristics include rigidity, toughness, impermeability to water and reagents, viscosity, adhesiveness. They provide the rigid skeleton of plants (cellulose, hemicellulose) and many invertebrates (chitin). Some act as adhesives, binding together the cells of plant tissue (pectin), while others may serve to cover and protect wounds in plants (gums). Some polysaccharides are the main constituents of the protective capsules or shields of many microorganisms.

(b) *"Nutrient" or "storage" polysaccharides* act as metabolic reserves (starch, glycogen). Monosaccharides in excess are condensed into polysaccharides and stored in this immobilized, insoluble form. The monoses are recovered from their reserve form by hydrolysis, whenever needed.

Lately, a number of polysaccharides have been found to possess antigenic properties.

As biopolymers, polysaccharides differ considerably from proteins, in that polysaccharides have much more uniform structures. As stated above, many contain a single monomer. Very often the glycosidic bond is uniform throughout the molecule. Thus β-(1→4) bonding is found in cellulose, almost exclusively. α-(1→4) bonding is essentially uniform throughout amylose. Even when the polysaccharide is branched, branching occurs at more or less regular intervals. This uniformity of structure is apparently due to the high specificity of the enzymes involved in the biosynthesis of carbohydrate polymers.

Any polysaccharide sample usually consists of a mixture of shorter molecules and longer ones. The distribution of molecular weights of *degrees of polymerization* (DP) about an average value is usually quite scattered. Part of this wide distribution is due to isolation techniques which degrade the native polymer to some extent. There are different ways to define "average" values for DP and molecular weight.

The *number-average molecular weight* M_n is defined as:

$$M_n = \frac{w}{\sum_{i=1}^{\infty} N_i} = \frac{\sum_{i=1}^{\infty} N_i M_i}{\sum_{i=1}^{\infty} N_i}$$

where w is the weight of the sample, N_i the number of molecules of a given polymeric species i in the sample and M_i the molecular weight of that polymeric species.

The *weight average molecular weight* M_w is given by:

$$M_w = \frac{\sum_{i=1}^{\infty} w_i M_i}{w} = \frac{\sum_{i=1}^{\infty} N_i M_i^2}{\sum_{i=1}^{\infty} N_i M_i}$$

where w_i is the weight of species i in the sample.

The corresponding average DP is obtained by dividing the average molecular weight by the molecular weight of the monomer residue.

The molecular weight of polysaccharides is determined by physical methods. Methods based on osmotic pressure and colligative properties are useful for polysaccharides which can be brought into solution without degradation. Colligative properties are, of course, a function of the number of molecules in a given sample and therefore give number-average values. Light scattering of dilute solutions and ultracentrifugation techniques give weight-average values. The viscosity of solutions also provide information on the molecular weight, but a different, third kind of average is involved.

X-ray diffraction measurements confirm the crystalline nature of most polysaccharides.

8-2. CELLULOSE

The structural material of the entire plant world consists largely of cellulose. In quantity, therefore, cellulose is the most abundant organic compound. It has been estimated that more than 50% of the carbon in the biosphere is bound in cellulose. A "tissue" consisting of several layers of cellulose fibers in parallel array is the main reinforcing element of the cell-wall in plants. Cellulose is the main component of many industrially important materials, such as wood, paper and fibers of plant origin. Cotton fiber is almost pure cellulose.

Chemically, cellulose is a linear polymer of D-glucose units, joined by β-(1→4) bonds, as in cellobiose:

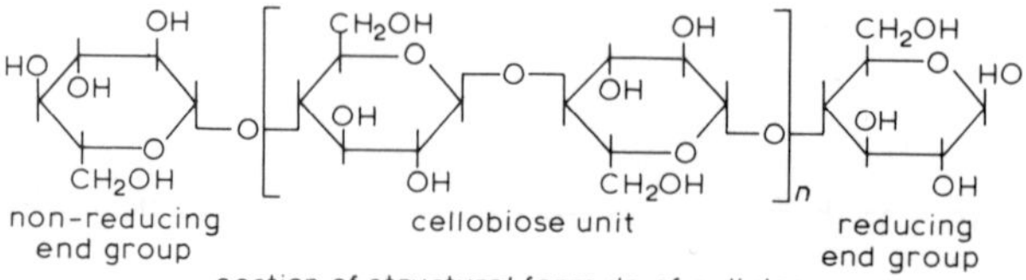

non-reducing end group — cellobiose unit (n) — reducing end group

section of structural formula of cellulose

The molecular weight of cellulose is very high. Values in the range of 10^6 have been reported. The molecular weight of native cellulose is difficult to measure due to the degradation which takes place during isolation. Apparently, the length of the chain varies from one plant to another.

In nature, cellulose occurs in the form of fibers. Cellulose fibers have very high mechanical strength. They are insoluble in water and highly resistant to chemical reactions. In fact, once formed, the cellulose fibers undergo very little change during the lifetime of plant tissue, unless attacked by fungi. The occasional occurrence of cellulose in fossilized plants, and the fact that paper and textile specimens have resisted degradation for thousands of years, are due to the remarkable chemical stability of the cellulose fiber.

The β-(1→4) glycosidic bond confers on the cellulose molecule a fairly extended, rigid configuration which approaches that of straight line segments. Furthermore, several chains can come very close one to another, without interference from the C-6 side group. Thus cellulose chains form bundles or microfibrils in which the chains run parallel one to another and to the axis of the microfibrils. These bundles are stabilized by abundant hydrogen bonds. In some regions of the microfibril, the chains align themselves in an even closer, regular pattern and form crystalline structures or crystallites. These "crystals" are too small to be observed with an optical microscope, but their presence is demonstrated by X-ray diffraction and other physical measurements. In summary, the cellulose microfibril is a bundle of cellulose molecules with "frozen" crystalline regions alternating with less orderly, amorphous regions where some mobility is possible. Even in the amorphous regions the chains are apparently packed together very closely, for the density of the dry cellulose fiber is very high and approaches that of the crystallite. The microfibrils are about 50 Å thick and 500–1000 Å long. The fiber consists of a bundle of fibrils parallel one to another but at an angle to the fiber axis.

The strong intermolecular forces that stabilize the microfibril structure also explain the mechanical strength and chemical stability of the cellulose fiber. In crystalline cellulose, many of the hydrophyllic groups are not available for hydration. In contact with water or water vapor, cellulose absorbs moisture and swells somewhat, but water is apparently adsorbed mainly at the free surface of the fibril and in its amorphous regions. The "frozen" crystalline regions still hold the bundle together, thus preventing its disintegration and dissolution.

Cellulose can be hydrolyzed into shorter chains, oligosaccharides, and finally, into cellobiose and glucose. Acid hydrolysis requires elevated temperatures and high acid concentration. Acid hydrolysis of wood and similar materials is the basis of an industrial process known as "*saccharification*". Enzymic hydrolysis is catalyzed by *cellulases*, which occur widely in fungi and some bacteria. Apparently different types of cellulases exist, which differ in their specific action. However, those types are not clearly defined as yet. Cellulase preparations are commercially available. Their use in food and related industries is still in the experimental stage. Suggested applications include enzymic saccharification of cellulosic materials, improvement of digestibility in animal feeds, improvement of rehydration rate in dehydrated vegetables, decomposition of cellulose–protein complexes in order to increase yields of protein extraction (e.g. in the isolation of protein from soybean oil meal). The industrial implementation of enzymic cellulolysis is probably retarded by the present high cost of cellulase preparations.

Man and most animals are unable to digest cellulose since they lack cellulases in their digestive tracts. The fact that ruminants are able to digest cellulose is due to the presence of cellulase-excreting microorganisms in the rumen. Similarly, termites can digest cellulose with the help of microorganisms living symbiotically in their digestive tract.

Cellulose can be modified chemically by substitution of the available hydroxyls in positions 2, 3 and 6. Esterification with nitric acid yields nitrocelluloses. Depending on the degree of esterification, these are used as explosives, adhesives or coatings. Cellulose acetates are important in the manufacture of transparent films, packaging materials, plastics and industrial membranes. Methylation and carboxymethylation yield soluble derivatives, used in food industries as thickening agents. These are also practically indigestible.

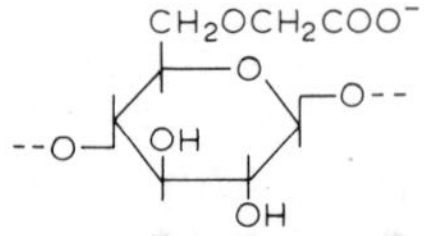

$CH_2OCH_2COO^-$

repeating unit of carboxy-methyl cellulose

"*Microcrystalline cellulose*" consists of very fine particles of cellulose, prepared by treating cellulose with dilute acid. This treatment apparently attacks the amorphous regions of cellulose, disintegrates the fibrils and leaves a colloidal suspension of the tiny crystallites. The use of microcrystalline cellulose as an indigestible non-caloric constituent in dietetic foods is increasing.

8-3. STARCH AND GLYCOGEN

8-3-1. *Occurrence and composition.* Starch is widely distributed in nature as a reserve material in almost every part of all plants. It contributes more calories to the normal diet of human beings than any other single substance. Its functional properties are important in many foods. Starch occurs in plant cells, in the form of insoluble particles, or *granules.* The size and shape of these vary from one plant to another and it is possible, by means of ordinary microscopic examination, to determine the source of a given starch (Fig. 17).

Total hydrolysis by acid or enzymes results in the quantitative conversion of starch of D-glucose. Thus, chemically, starch is a *glucan* or rather a mixture of glucans, for the starch granule is a heterogenous system consisting mainly of two different components:

amylose, which is an essentially linear polymer;
amylopectin, which is a highly branched polymer.

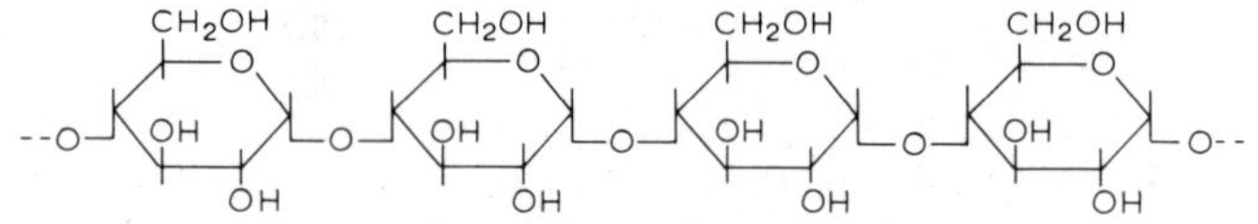

portion of the amylose molecule

portion of the amylopectin molecule

Depending on the source, the amylose molecule contains 1000 to 5000 monose residues linked together mainly through α-(1→4) bonds. However, "imperfections" such as branching or other types of glucosidic bonds are found along the chain with considerable frequency.

Amylopectin, on the other hand, is a much larger molecule, in fact so large that accurate determination of its molecular weight becomes extremely difficult. Only estimates are available and these place the degree of polymerization of amylopectin in the range of 10^5–10^6 monose residues per molecule. Methylation and subsequent hydro-

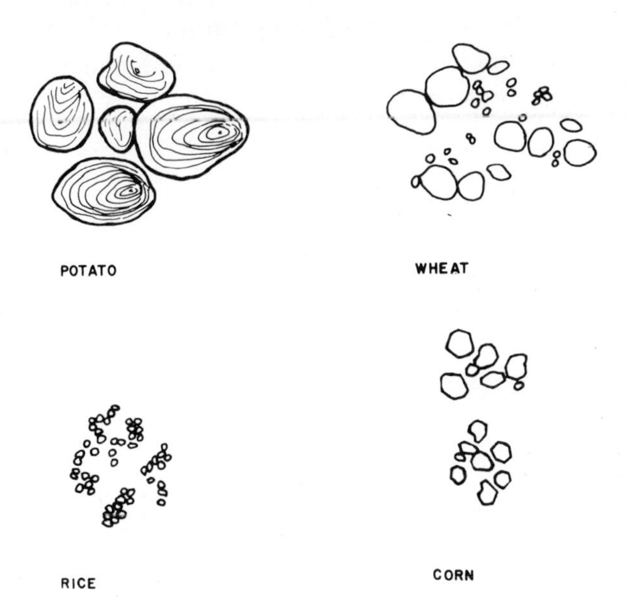

Fig. 17. Characteristic shapes of various starch granules (magnification approx. x 300).

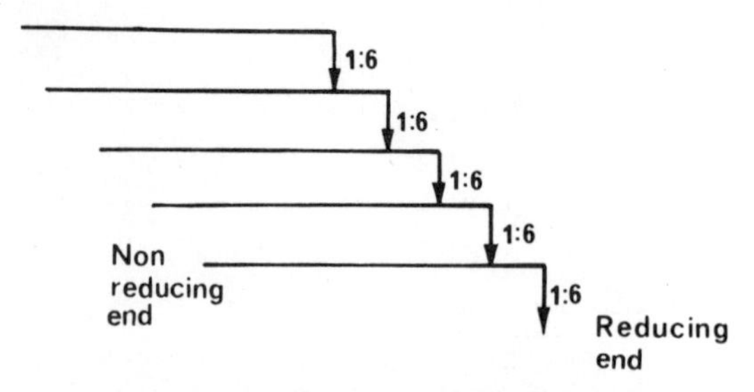

Fig. 18. The Haworth model of amylopectin.

lysis yields mainly 2-3-6 trimethyl and 2-3 dimethyl glucose, indicating that amylopectin consists of short amylose chain segments, interconnected through α-(1→6) bonds. The chain segments contain 20 to 30 glucose residues each. Other bonds, mainly α-(1→3) linkages, are also found, but much less frequently. The exact pattern of branching is not known but several models have been suggested, mostly on the basis of physical properties of colloidal amylopectin solutions. The Haworth model, shown in Fig. 18, consists of a laminated pattern in which each amylose chain is attached to the next at one branching point, through α-(1→6) link, so that each chain has one branching point. The Staudinger model (Fig. 19), consists of a main chain of amylose to which unbranched

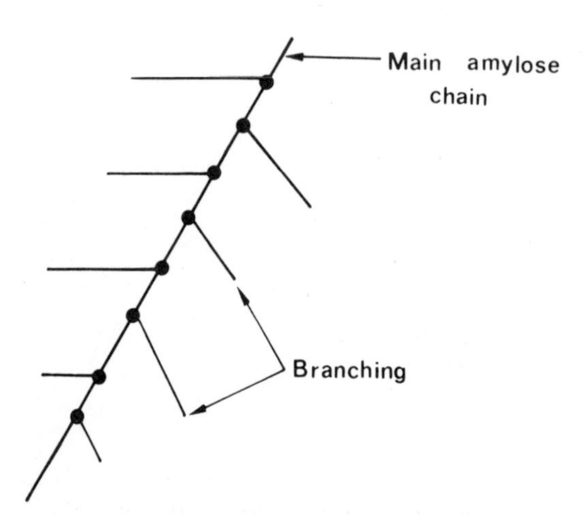

Fig. 19. The Staudinger model of amylopectin.

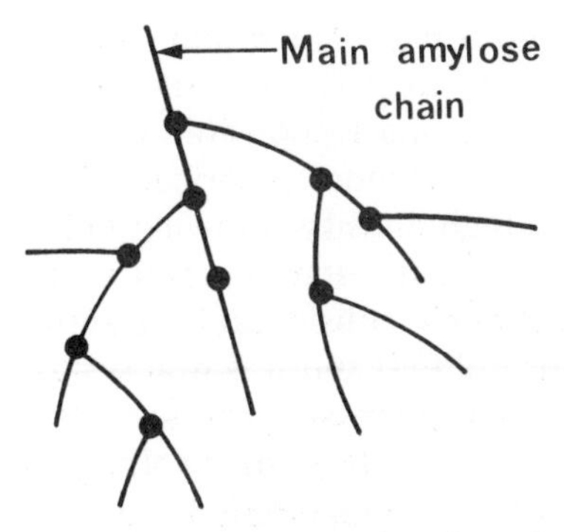

Fig. 20. The Meyer model of amylopectin.

side chains are attached by α-(1→6) bonds. The Meyer model (Fig. 20), is a completely ramified structure combining the features of the two previous models. This latter is considered as the most plausible model.

The relative proportion of amylose and amylopectin varies from one starch to another. Usually, starches contain more amylopectin than amylose. The amylose to amylopectin ratio is governed by genetic factors and can be influenced by plant breeding techniques. Thus, while in ordinary corn starch this ratio is approximately 1 : 3, starch from waxy corn varieties contain almost no amylose. On the other hand, high-amylose corn, a product of modern plant breeding, may contain in its starch up to 80% amylose.

8-3-2. *Structure and properties of the starch granule.* The starch granule usually consists of layers, laid down around a central region called the *hilum.* The origin of these layers is not known. They have been explained by diurnal alternation in the synthesis of starch, but this view seems to be incorrect. Some degree of order in the arrangement of the molecules in the granule is evident from X-ray diffraction. These patterns are believed to be due to the presence of microcrystallites consisting mainly of the short side chains of the amylopectin component. They are not observed in high amylose starches. Partial hydrolysis with cold acid produces microcrystalline particles, just as in the case of the cellulose fiber. The proportion of crystallites in the starch granule is, however, lower than that in the cellulose fiber and the starch crystallites do not possess the high mechanical strength of their cellulose counterpart. The orientation of the crystallites in the starch granule seems to be parallel to the radius of the granule. This structure is believed to be responsible for the *birefringence* of the starch grain when viewed under polarized light. Starch in the granular native form has a high density, showing fairly close packing within the granule. Therefore, the starch granule can withstand some mechanical action (although considerable granule damage may occur during industrial flour-milling) and it is practically insoluble in cold water. If the temperature is increased, however, the granule adsorbs water and swells. The initial stage of swelling is reversible and the granule may be dried back to its original state. As the temperature is increased further, swelling continues, birefringence becomes less clear and finally disappears. The crystallites have "melted". At this point, the starch is said to have reached "*gelatinization*". Finally, the granule disintegrates and starch goes into solution. These final stages are irreversible. If a freshly prepared starch paste is dehydrated, an amorphous powder, readily dispersible in cold water, is obtained. This is the basis of the manufacture of "*soluble starch*" and "instant" precooked cereal products.

8-3-3. *Properties of amylose and amylopectin.* The α-(1→4) glucosidic bond is fairly flexible. The shape of amylose molecules in aqueous solution is primarily that of random coils. Parts of the chain may consist of short, unstable, transient helical segments. The helical form is stabilized by several complexing agents, such as iodine. Each turn of the helix comprises six hexose units and the segments may

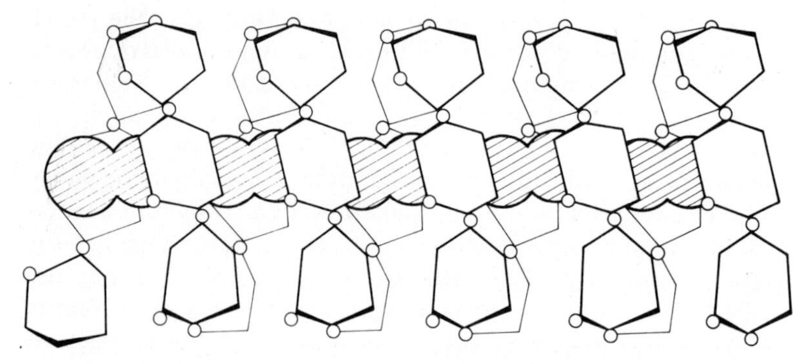

Fig. 21. Helical model of amylose-iodine complex (shaded areas represent iodine).

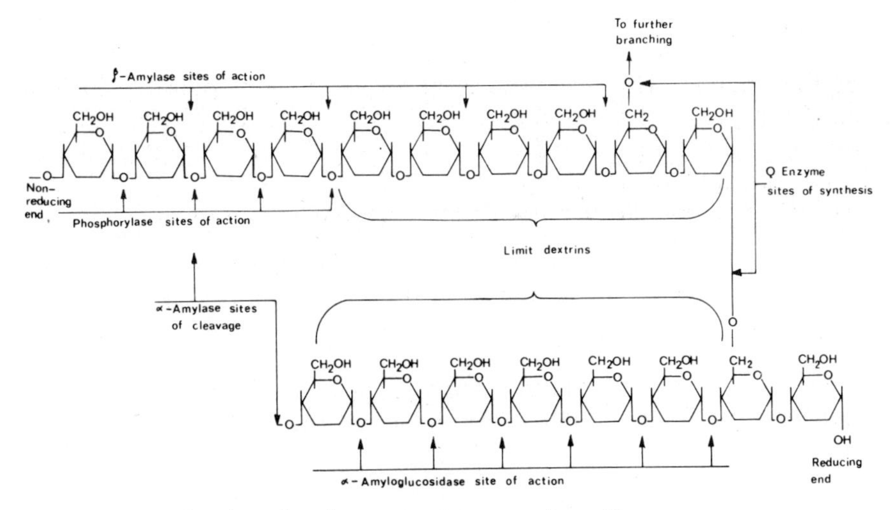

Fig. 22. Sites of action of various enzymes on amylopectin.

have 4 to 20 turns. One of the best known characteristics of amylose is its ability to bind *iodine*, to give a deep blue complex. The reaction is a familiar quantitative test for starch. Amylose can bind up to 20% iodine by weight. Iodine is adsorbed inside the amylose helix and stabilizes the helical conformation (Fig. 21). If the complex is heated the blue color disappears, but returns on cooling. Short chains of amylose obtained by partial hydrolysis (dextrins) adsorb little iodine and give a reddish complex.

When a solution of amylose is cooled, a gel is obtained. Starch gels are responsible for a good part of the structure in custards, sauces and many dough products and cereals. They are formed by thermoreversible "fusion" of amylose chains at certain points, while the remainder of the chain retains its random configuration. The points of fusion between chains are most probably hydrogen bonds formed at regions where the chains come sufficiently close together. The solvent (water in the case of foods) is immobilized within the loose, tridimensional structure formed by the chains. However, if a solution of amylose is cooled very slowly, or if an amylose gel is aged, the amylose molecules align themselves in a more orderly fashion, forming dense, highly crystalline, insoluble micelles. This process is named "*retrogradation*".

Amylopectin has a high intrinsic viscosity, as can be expected from its high molecular weight and branched structure. Sols of amylopectin-rich starches, such as waxy corn starch, are thick pastes. Solidification of such pastes into a stiff gel is a very slow process, requiring a few days. Amylopectin retrogrades very slowly, if at all. In a mixture of amylose and amylopectin, retrogradation of amylose is inhibited by the presence of amylopectin. Neverthleless, retrogradation of the short side chains of amylopectin has been suggested as the main factor in the *staling of bread.* Amylopectin adsorbs only a small amount of iodine and gives a red color complex. The average iodine binding capacity of a starch sample is a good indication of the amylose–amylopectin ratio. Amylopectin contains usually some phosphorus (0.02 to 0.05%), probably in the form of ester linked phosphate.

8-3-4. *Glycogen. Glycogen,* the storage polysaccharide of the animal kingdom, resembles amylopectin in composition and structure. It is made of glucose residues, linked by α-(1→4) bonds, the resulting chain segments being held together by α-(1→6) linkage at branching points. The average chain segment length is approximately 12 monose units. The molecular weight is very large, probably in the range of 10^8. Glycogen "particles" can be observed by electron microscopy. Considering the molecular weight, it is not unlikely that these particles are in fact single molecules or clusters consisting of a few molecules at most.

8-3-5. *Starch enzymes.* In order to make the polysaccharides, such as starch, available for metabolic transformations, they must be broken up into smaller units, i.e. their glycosidic linkages must be split. A number of enzymes are known to catalyze the breakdown of starches and some other enzymes are capable of catalyzing their biosynthesis. In general, the action of starch splitting enzymes on *whole starch granules* is extremely slow. Nevertheless, amylolysis in the whole grain is important in nature as it represents the in vivo mechanism for the liberation of metabolizable carbohydrate from the storage polysaccharide in plants. Enzymic breakdown of gelatinized starch is, of course, much more rapid. This explains the need for cooking or roasting of starch-rich foods (cereals, potatoes) for better digestibility.

The splitting of the glycosidic bond may be represented schematically as follows:

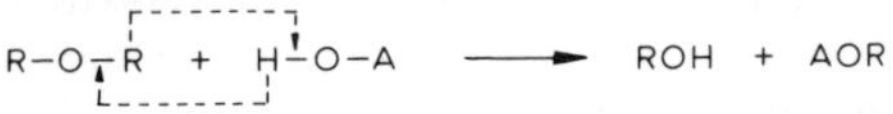

$$R{-}O{-}R + H{-}O{-}A \longrightarrow ROH + AOR$$

If H—OA is water, we have a case of *hydrolysis*, catalyzed by *hydrolases*. Frequently in nature the H donor molecule is phosphoric acid rather than water. In this case the breakdown of the glycosidic bond will produce one molecule of unsubstituted carbohydrate and another becoming a phosphate group. The reaction known as *phosphorolysis* is catalyzed by *phosphorylases*.

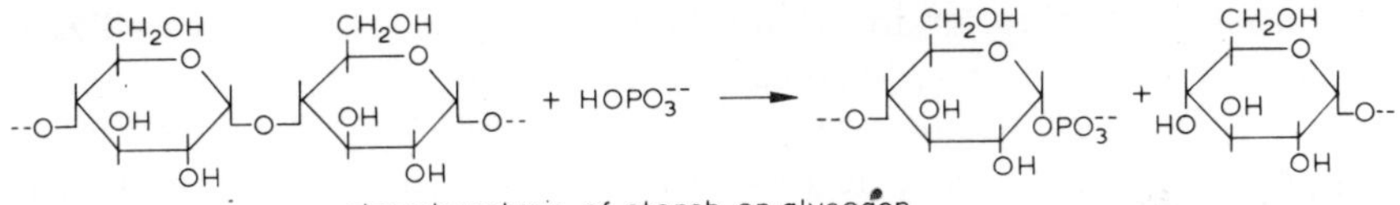

$$\cdots + HOPO_3^{--} \longrightarrow \cdots OPO_3^{--} + HO \cdots$$

phosphorolysis of starch or glycogen

Hydrolases which split the α-(1→4) bonds of starch are known by their trivial names as amylases. Amylases are of two main types: *α-amylases* and *β-amylases*. The prefixes α and β refer only to the mutarotation of the hydrolysis products, and have no connection with the bond attacked.

α-Amylases are widely distributed in nature. They are the starch digesting enzymes of saliva and the pancreatic excretions in animals. They are also present in plants and microorganisms. α-Amylases from several sources have been crystallized. α-Amylases are metallo-enzymes, containing bivalent metal, usually calcium or zinc. Their optimal pH range is about 5 to 6. When α-amylase reacts on an amylose sol, a rapid decrease in viscosity with little increase in concentration of reducing groups occurs. Thus, α-amylase is an *endo-amylase* attacking the polysaccharide at glycosidic bonds *inside* the chain. Therefore, the first product of α-amylase catalyzed hydrolysis of amylose is a mixture of shorter amyloses. As hydrolysis proceeds α-amylase liberates maltose and some D-glucose. α-amylosis of long chains is a random process, but some preference as to the site of hydrolysis may exist when the substrate contains shorter chains. When amylose is subjected to the prolonged action of α-amylase, most of it is transformed into glucose and maltose, but a small proportion remains as a mixture of short polysaccharides which are not hydrolyzed further. This mixture is named *α-limit dextrin of amylose.* The inability of α-amylose to hydrolyze this type of limit dextrin is explained by the presence of "barriers" to α-amylolysis in amylose, such as branching, oxidized hexose residues or bonds other than α-(1→4).

α-amylase hydrolyzes amylopectin starting with the outer chains but breaking also α-(1→4) bonds between branching points. The action stops, however, a few residues away from a branching point. Thus, the products of α-amylolysis of amylopectin are glucose, maltose and branched oligosaccharides containing five or more monose residues.

β-amylase is found almost exclusively in higher plants. Most convenient sources for its isolation are the sweet potato and the germinating barley grain. β-amylases have also been crystallized. β-amylase is an *exo-hydrolase* specific towards the penultimate α-(1→4) bond of the non-reducing end of the chain. The product of β-amylolysis of amylose is primarily maltose but also some glucose resulting from the reducing end of chains containing an odd number of residues, and some dextrins due to "barriers" as discussed above. In the case of amylopectin, the action of β-amylase is stopped in the vicinity of a branching point. Thus β-amylase hydrolyzes only the outer chains of amylopectin, leaving a large core of polysaccharide delimited by branching points. This residue is named *β-amylase limit dextrin* of amylopectin. β-amylases are not known to require any co-factor.

Amyloglucosidase is another starch splitting enzyme, found mainly in molds. As β-amylase, this is an *exo-hydrolase* but it attacks the last glucosidic bond at the non-reducing end of the chain, liberating glucose rather than maltose. Unlike the maltases, amyloglucosidases hydrolyze not only α-(1→4) bonds but also α-(1→6) and α-(1→3) linkages, although at a much slower rate. Therefore, amyloglucosidases are capable of unbranching limit dextrins. They can hydrolyze small glucans and even maltose, but the rate of reaction is many times faster in the case of larger polysaccharides.

Pullulanase is a glucosidase specific towards α-(1→6) bonds. It has been isolated from *Aerobacter aerogenes* and studied extensively in the last decade. It derives its name from its action on *pullulan*, a polymer of 1→6 linked α-maltotriose residues. Its specificity to 1→6 bonds is the basis for its use in the study of the structure of amylopectin and glycogen.

Phosphorylase, which is capable of detaching single molecules of glucose from the non-reducing end of amylose chains and of phosphorylating them simultaneously at the C-1 position, is widely distributed, both in animals and plants. It has been isolated from potatoes, pea seeds, liver and other sources. This important enzyme is responsible for the liberation of glucose from glycogen in the liver and muscle. Its activity in the animal tissue is regulated by an elaborate control system.

The natural transformation of glucose to starch and vice versa is a process of utmost technological importance in the case of many fruits, vegetables and cereal grains. As was mentioned before, starch is the storage carbohydrate of plants. In the process of maturation, starch accumulates in tubers, roots and seeds. To the technologist this transformation is sometimes undesirable, as in the case of sweet corn and peas which are more acceptable when they are immature and sweet. On the other hand, potatoes for baking and frying are preferred starchy and less sweet, therefore mature. Transformation of sugars to starch may take place after harvest as it is experienced with immature potatoes or peas, stored above 20° C. In cold storage the process is reversed. When potatoes are stored at 0°C, some starch is degraded and sugars, mainly sucrose and glucose, accumulate rapidly. The main objection to high sugar content in potatoes is the tendency of sugar-rich tubers to undergo browning when fried. Many fruits (apples, bananas) accumulate starch during growth. As maximum growth is attained and ripening begins, starch is converted to sugars and the fruit becomes sweet. Often, these changes can be made to occur in post-harvest artificial ripening.

"Intentional" starch hydrolysis by added enzyme preparations is an old technological process. The art of preparing *malt* (malting) is very old. Barley (sometimes wheat) is moistened to 45-50% moisture content, then spread on the floor and allowed to germinate. During the germination period which lasts approximately one week, the amylolytic activity increases rapidly. Germination is stopped by drying the grains in a kiln. The product, malt, has a brown color and characteristic flavor, but it is mainly used for its amylolytic activity. It contains both α- and β-amylases. While ungerminated grains have considerable β-amylase activity, all of the α-amylase is produced during germination.

Certain molds, and particularly *Aspergillus oryzae*, are extensively used in the Far East for their amylolytic properties. Today, purified amylase preparations are obtained commercially from various fungi

(*Aspergillus oryzae*, *A. niger*, *Rhizopus* species) and bacteria (mainly *Bacillus subtilis*). In virtue of their specific properties these preparations find interesting applications in many fields of tood technology.

(a) *Baking.* It was an old practice in the baking industry to supplement wheat flours with malt. In bread dough, the presence of fermentable sugars is essential for the rapid growth of yeast and adequate gas production. The free sugar content of flour from ungerminated wheat is usually too low. Cereal flours contain considerable β-amylase activity but the absence of α-amylase limits the rate of maltose formation. Amylolysis by the enzymes of the yeast is not sufficiently rapid. Thus, supplementation with amyloses from external sources is beneficial, as it accelerates fermentation and improves gas production especially in the case of flours deficient in amylolytic activity. Due to their α-amylase content these additives have an immediate "liquefying" action which reduces the viscosity of the dough. This is often desirable. α-amylases from bacterial sources are relatively heat resistant and are not destroyed completely when the bread is baked.

It has been mentioned before that whole starch granules are resistant to enzymic hydrolysis. The action of amylases in the dough is limited mainly to the "damaged" portion of the starch. During baking, a considerable percentage of the starch undergoes gelatinization and becomes susceptible to hydrolysis by the residual activity of heat resistant α-amylases. α-amylolysis in the oven, in the initial stages of baking, is known to result in softer crumb texture, which remains soft for many days after baking.

(b) *Industrial hydrolysis of starch.* The conventional process for the industrial hydrolysis of starch uses strong acids and high temperatures. Dextrins are formed first, which are gradually broken down to oligosaccharides, maltose and finally glucose. The resulting water soluble mixture is concentrated into syrups. As corn is the main source of starch for this purpose, the product is known as *corn syrup.* Depending on the extent of hydrolysis, corn syrups may contain different proportions of dextrins, maltose and glucose. Their viscosity, sweetness, flavor and many of their functional properties depend on the extent of hydrolysis, which is expressed as their *dextrose equivalent* (DE). The dextrose equivalent is the reducing sugar content (as glucose) expressed as a percentage of the total soluble solids. Corn syrups with different DE values find many uses in the food industry as texture conditioners, sweeteners, carriers, crystallization inhibitors, etc.

If hydrolysis is carried far enough, practically all the carbohydrate is transformed to glucose which may be recovered as a concentrated syrup or as a crystalline solid.

Acid hydrolysis requires expensive corrosion resistant equipment. The extent of hydrolysis is difficult to control, especially towards the end of the process. Some furfural is formed and causes some browning and caramelized flavors. Enzymic hydrolysis as an alternative industrial process has been investigated for a long time and is now practised extensively. The process usually consists of two steps: *liquefaction* and *conversion*. The purpose of the first stage is to reduce the viscosity of the gelatinized starch sol so that more concentrated slurries may be used. This stage may be carried out by bacterial α-amylases above 80° C. For the second stage of saccharification or conversion fungal glucoamylases are used, if the final product has to be glucose. Corn syrups with varying dextrin maltose/glucose ratios can be manufactured by using different mixtures of α,β-amylases and glucoamylases.

(c) *Fermentation industries.* Starch-containing agricultural commodities such as cereal grains and potatoes constitute a major source of carbohydrate for the industrial manufacture of alcohol and alcoholic beverages. The grains are cooked (mashed) and treated with malt to convert the gelatinized starch to fermentable sugar. Fungal amylases are now replacing malt in the manufacture of "neutral spirits" from starch-containing raw materials. In the brewing industry, malt is still the principal saccharifying agent, mainly because of the contribution of malt to the characteristic flavor and color of beer.

8-3-6. *Manufacture of starch.* Industrially, starch is produced either from root plants (such as potatoes or sweet potatoes) or from cereals (wheat, corn, rice, etc.). Corn is the principal source of industrial starch.

About 16 to 22% starch can usually be obtained from potatoes. By genetic improvement of certain varieties it has been possible to obtain a yield of 25 to 40%. The main manufacturing steps comprise:

(1) Thorough washing to remove earth and impurities.
(2) Complete disintegration to open all cells.

(3) Washing out the starch granules and separating them from the pulp upon vibrating screens.

(4) Addition of SO_2 to prevent browning.

(5) "Tabling" — separation of the soluble impurities from the starch milk by passing it through long channels in which the starch settles. This step can also be achieved by centrifugation or hydrocyclones.

(6) Washing, bleaching with SO_2 and permanganate.

(7) Drying with warm air at 30° to 40° C.

In manufacturing starch from cereals the above steps must be preceded by removal of the proteins before the precipitation of the starch. Consequently, corn grains, for instance, are steeped in water containing 0.3% SO_2 for 30 to 40 hours in order to soften them. They are then subjected to wet milling in order to remove the germ by flotation. The germ is dried and sent to oil extraction plants for the recovery of corn oil. Subsequent steps for the extraction of starch are similar to those used for root materials. Removal of proteins from wheat is done by working the dough and separating the gluten mechanically. In this so-called Alsace process, one gets 500 kg of first-grade starch from a ton of flour and an additional 200 kg of second-grade starch.

8-4. HEMICELLULOSES

Under this generic name we designate a large number of complex polysaccharides which accompany cellulose in the cell wall of plants. Most hemicelluloses are heteropolysaccharides containing two or more different monose residues. Some hemicelluloses consist of a homoglycan main chain (D-xylans and D-mannans) with other sugar residues occurring mainly as side chains. Uronic acids (mainly D-glucuronic acid) participate frequently in the composition of the polymer, and therefore, hemicelluloses show acidic properties. Acetyl groups are also common. The residues of the main chain are usually linked by α-(1→4) glycosidic bonds. Due to their heterogeneity and the presence of bulky side groups, hemicelluloses are much less crystalline and therefore much more soluble than cellulose. Hemicelluloses are easily extracted from fibers and wood by alkaline solutions. Hemicellulose hydrolyzing enzymes are widely distributed in nature. Obviously, a very large number of different enzymes participate in the hydrolysis of so heterogeneous a group such as hemicelluloses.

8-5. OTHER POLYSACCHARIDES

Chitin is an important polysaccharide which constitutes the major structural part of invertebrates. It is also found in large quantities in the shells of crustaceans and some molluscs. It is also found in structural tissues in most fungi and algae. Chemically, chitin is a linear poly-N-acetyl-D-glucosamine. The glycosidic bond is β-(1→4). Chitin is very similar to cellulose in structure and function. It is fibrillar and highly crystalline.

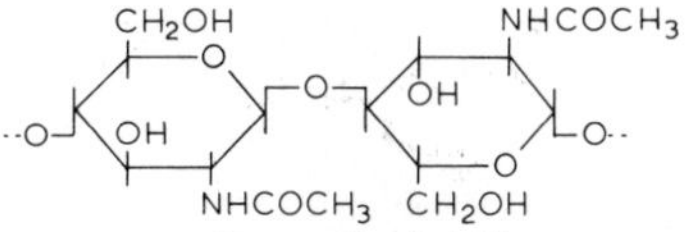

CH_2OH $NHCOCH_3$ OH OH $NHCOCH_3$ CH_2OH

repeating unit of chitin

Chitin is the most abundant polymer of amino sugars. Other important polysaccharides containing nitrogen are constituents of the bacterial cell wall and the *mucopolysaccharides* of higher animals. These latter occur in the extracellular body fluids and are believed to fulfil important biological functions as mechanical lubricants, tissue binders and regulators of interstitial permeability. The most important members of the group (*hyaluronic acid*, *chondroitin*, etc.) have been the subject of intensive research in medicine.

A number of microorganisms are able to synthesize peculiar polysaccharides outside their cells, by means of extracellular enzymes. One such polysaccharide, *dextran*, is produced commercially. It has been used as a blood plasma substitute for transfusions. Dextran consists of D-glucose residues linked mainly by 1→6 bonds. A similar polymer of fructose (*levan*), is also produced by a number of microorganisms.

Pectins and plant gums are also polysaccharides. In view of their special importance in foods, these substances will be discussed separately in the next chapter.

SELECTED BIBLIOGRAPHY TO CHAPTER 8

Greenwood, C.T., Structure, Properties and Amylolytic Degradation of Starch, J. Food Technology, **18** (1964), 732-738.

Meyer, K.H., Natural and Synthetic High Polymers, 2nd edn., Interscience Publishers, New York, 1950.

Pigman, W. and Horton, D. (Eds.), The Carbohydrates, Chemistry and Biochemistry, 2nd edn., Vols. IIA and IIB, Academic Press, New York, 1970.
Radley, J.A., Starch and its Derivates, 4th edn., Chapman and Hall, Ltd., London, 1968.
Sterling, C., The Submicroscopic Structure of the Starch Grain. An Analysis, J. Food Technology, 19 (1965), 987-990.
Wurtzburg, O.B. and Szymanski, C.D., Modified Starches for the Food Industry, J. Agr. Fd. Chem. 18 (1970), 997.

CHAPTER 9

PECTIC SUBSTANCES, PLANT GUMS

9-1. OCCURRENCE

A very important group of substances known in food technology as *pectins* belong to the second group of polysaccharides, the heteropolysaccharides. The name pectin, from the Greek adjective πηχτοϛ (curdling, stiff) was given to these substances by Braconnot in 1825, in recognition of their property of forming gels. It is, in fact, only a generic name embodying a group of closely related substances — the pectic substances. These fill the intercellular spaces, the middle lamella, of plant tissues. In young tissues, especially in fruits, pectins are often present in such large amounts that they form wide channels, pushing apart the cells (Fig. 23). Being a hydrophillic colloid, pectin has the property of imbibing large quantities of water. Because of this capacity, pectic substances apparently play an important role in the early stages of development of plant tissues when the cells still lie apart and at a comparatively great distance

Fig. 23. Diagrammatic drawing of pectin in channels as seen under the microscope.

from the water-conducting vessels. The pectic substances quickly absorb water and transfer it among the cells more easily than could be effected by osmosis in the cells themselves. As a natural constituent of plant tissue, pectic substances are responsible for a great part of the firmness and texture of fruits and vegetables. Softening of fruit tissue during ripening, breakdown of colloidal stability in fruit juices, changes in the consistency of fruit purees and concentrates can often be attributed to changes in pectic substances. As intentional additives, pectins are valuable gel forming and thickening agents.

As long as pectic substances are in the outer cell walls in the region of the middle lamella in the plant, they are thought to be closely associated with cellulose. In this form, as a precursor of pectin proper, the substance is called *protopectin.* Protopectin is insoluble in water and, according to Branfoot (Carré), can be observed microscopically in the plant tissue using ruthenium red as a stain. The exact nature of the association between pectic substances and other cell wall constituents is not known. Joslyn mentions several possibilities ranging from molecular cohesion to actual covalent bonding. Association between carboxyl groups has been postulated by Henglein.

When pectin-rich plant materials, such as apple pomace or citrus peels, are heated with acidified water, the protopectin is liberated, probably from the adhering cellulose, and transformed to water soluble pectin. The same transformation of the protopectin takes place in plant tissues during ripening of the fruits apparently with the aid of an unknown enzyme to which the name *protopectinase* has been assigned. Many pectin chemists question the existence of such an enzyme. The liberation of soluble pectin and softening of tissues may well be due to enzymic degradation of cellulose and hemicellulose rather than to depolymerization of protopectin. Until a concensus is reached on this matter on the basis of conclusive evidence it may be advisable to adopt the nonspecific name "*macerase*" or *macerating enzyme* instead of protopectinase.

Pectin is readily precipitated from aqueous solutions by alcohol or acetone as a jelly-like coagulum, which will in turn again dissolve in water. This coagulation can also be brought about by the action of a mixture of certain salts, such as aluminium sulfate in conjunction with ammonium hydroxide, whereby aluminium hydroxide is formed, the colloidal particles of which carry an electric charge of opposite sign to that of pectin (pectin is a negatively charged colloid).

9-2. STRUCTURE

Pectin is, therefore, a reversible hydrophilic colloid; its solutions rotate polarized light to the right. Crude commercial pectin contains a number of impurities, such as hemicelluloses, pentosans (araban), galactosans and other compounds, but can be purified by repeated precipitation and redissolution.

Chemically, pectin consists of long, unbranched chains of *poly-galacturonic acid*, with the carboxyl groups partially esterified with *methyl alcohol*. The link between the galacturonic acid units is (1→4). The molecular weight varies from 20,000 to over 400,000. Neutral sugars and other uronides in proportions ranging from a few per cent to up to 20% are frequently found in pectin preparations. Most probably, these are not present as impurities, but as side chains or even as constituents of the main chain.

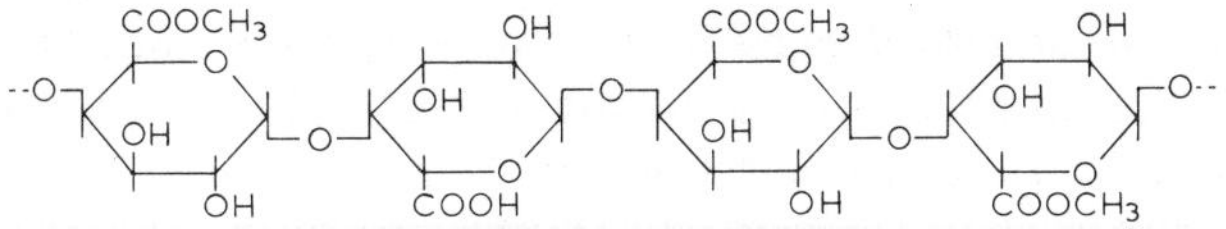

portion of pectin molecule

Pectins derived from different sources vary widely in their jelly forming properties due to the different lengths of their poly-galacturonic acid chains and to the different degree of esterification of their carboxyl groups with methyl alcohol. These also vary greatly according to the different methods of extraction employed. There are probably no two pectin preparations which are identical as far as their structure is concerned.

Pectins may undergo hydrolysis by acid or alkali or by the action of suitable enzymes. The first step in such hydrolysis is the removal of a varying number of methoxyl groups leaving ultimately poly-galaturonic acid, called also *pectic acid*, completely free of methoxyl groups. The many intermediate components, still possessing a varying number of methoxyl groups, give rise to a large number of *pectinic acids*. This is why pectin can be regarded only as a generic name covering a wide range of pectinic acids differing in their degree of polymerization and in their degree of esterification. Fully esterified pectic acid has never been found in nature; however, it may be synthesized for research purposes. Complete hydrolysis of pectic acid (polygalacturonic acid) results in the formation of single D-galacturonic acid units.

Some of the OH groups on C_2 and C_3 of the galacturonic acid units may occur in the acetylated form (sugar beet pectin).

9-3. PECTOLYTIC ENZYMES

A number of enzymes have been found to catalyze the various stages of pectin breakdown:

Pectin esterase (PE) catalyzes the hydrolytic removal of the methoxyl groups (saponification) from the pectin molecule. (Synonyms: *pectase, pectinmethoxylase, pectin methyl esterase.*) This enzyme is not always present in fruits rich in pectin. While tomatoes and all citrus fruits contain PE in abundance, this enzyme has not been found in beets, carrots, and is found in only a few varieties of apples. Most fungi contain appreciable amounts of PE. So far, pectin esterase has not been found in yeasts. The optimum activity of fruit esterase occurs at pH 7.5 and the enzyme is found to be quite specific: it requires the presence of a free carboxyl group next to the methoxyl group to be saponified, as shown by the following scheme. With the diminution of the number of such molecular arrangements the PE activity will decrease.

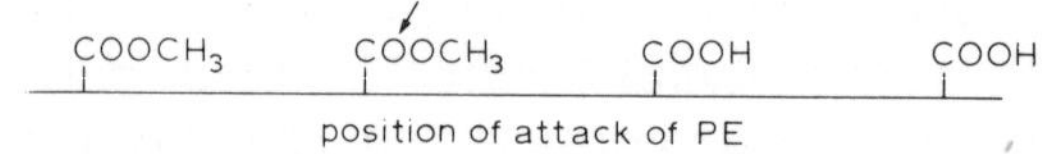

It has been demonstrated that in some cases PE is associated with solid particles of the cell wall. Such an association has been found by Bonner and his coworkers in orange juice where PE is a structure-bound insoluble enzyme. Such de-esterified cell walls have been shown to have a strong affinity for PE: approximately 5 mg of this enzyme can be bound by 1 mg of pectic material in the cell wall. The activity of this bound enzyme acting on cell wall pectin has been compared with that of the soluble enzyme acting on soluble pectin. Both have optima at pH 7.5. However, while the bound enzyme is inactive at pH 4.5 and below, the soluble PE is still active in this pH region. According to Jansen et al., the adsorption of orange PE on orange pulp is due to the formation of an enzyme-substrate complex. One of the practical implications of these findings, as far as orange juice is concerned, is that juices rich in pulp have higher PE activity.

Polygalacturonase (PG) catalyzes the glycosidic hydrolysis of the bond between galacturonic acid units. (Synonyms: *pectinase, pectolase, polygalacturonidase, pectin depolymerase.*)

In recent years various investigators have shown the existence of several different types of pectin depolymerizing enzymes. Some will cause rapid decrease in the viscosity of pectin solutions without considerable increase in reducing groups. These "*liquefying*" enzymes attack the molecule at random, breaking it to shorter chains. Because the site of attack is inside the chain these are also named "*endo-PG's*". Other enzyme preparations, "*saccharifying PG's*" or "*exo-PG's*", will split preferentially terminal galacturonic acid units and therefore increase the concentration of reducing groups before any substantial depression of viscosity occurs.

Pectin depolymerases may also differ in their specificity towards the state of methoxylation about the point of attack. Most common depolymerases of plant origin split glycosidic bonds between two unesterified units. Their substrate is polygalacturonic acid and the name *polygalacturonase* (PG) is in order for this type of enzyme. PG is inactive towards fully methoxylated polymer. However, in the presence of PE, which is capable of demethoxylating the pectin, PG can split all the links of the chain and produce free galacturonic acid. On the other hand, a group of fungal depolymerases will attack preferentially pectins with a high degree of methoxylation. These enzymes are named *polymethylgalacturonases.*

Phaff and Demain presented a summary of the different types of hydrolytic enzymes depolymerizing pectins, and a suggestion for a system of nomenclature. The authors also point out that endopolygalacturonase of yeast is a single enzyme. The presence of at least two types of polygalacturonase in tomato has been demonstrated.

As far as protopectinase is concerned, as explained above it is now doubtful whether a special enzyme, the so-called "protopectinase", exists or is altogether necessary.

Almost all fungi contain the full spectrum of pectolytic enzymes. This can be easily observed in different natural phenomena when the plant tissue is entirely disintegrated and falls apart, e.g. when some molds attack it; a fruit attacked by *Penicillum glaucum*, for instance, will ultimately disintegrate completely, since the pectin which has held together the fruit cells has been converted by the enzymes of the fungi into pectic acids which have no more binding power. A similar phenomenon can be observed during the retting process of flax or the fermentation of tobacco leaves and of coffee berries.

Industrial preparations of mixed pectolytic enzymes are manufactured from fungi and sold under the trade-names such as "*Pectinol*" and "*Mylase*".

Pectin transeliminase (PTE) also catalyzes the depolymerization of pectin. However, unlike PG, PTE does not hydrolyze the glycosidic bond, but actually eliminates it without the addition of a molecule of water as follows:

PTE was discovered by Albersheim, Neukom and Deuel, in an enzyme preparation of fungal origin. Its presence in fruit has not been reported. Its optimum pH is 5.1 to 5.3. The enzyme attacks only glycosidic bonds between methoxylated units. Incidentally, cleavage of the glycosidic bond in pectin by alkaline solutions has been reported to follow the same mechanism, with the formation of an unsaturated bond between C_4 and C_5 (Neukom and Deuel).

9-4. PECTIN AS A JELLIFYING AGENT

The most important use of pectin in food is based on its ability to form gels and it is, therefore, widely used in the manufacture of jams, jellies, marmalades, and preserves. In order that a pectin may form a gel, a dehydrating agent must be present: alcohol or acetone are typical dehydrating agents used for precipitation of pectin from its solutions. In the production of jams and jellies, it is the sugar that plays this dehydrating role. In forming a good jelly, suitable pectin–acid–sugar ratio should be maintained. The practical results of numerous researches in this field show that it is best to adjust the amount of pectin and the acidity in such a manner as to save sugar. An increase in acidity from 0.1 to 1.7% results in a saving of nearly 20% sugar. The same is true of pectin. Within certain limits (0.5–1.5% pectin content) the higher the percentage of pectin in the fruit pulp or juice, the lower the amount of sugar required to form a jelly.

Comprehensive investigations made by Tarr have shown that there is no direct relation between total acidity and jelly formation; only a strict limit of pH values should be maintained. This definite limit of hydrogen-ion concentration is found by Tarr to be at a pH value of 3.46, when a soft delicate jelly is given. Jelly increases in stability at lower pH values (down to 3.1–3.2); at still lower pH values syneresis occurs, i.e. the jelly exudes liquid on standing.

As far as pectin is concerned, some fruits rich in pectin, such as apples, quinces and citrus fruits, have excellent jellying power of their own. To jams made of other fruits poor in pectin, such as strawberries and many other berries, commercial pectin has to be added. The number of methoxyl groups present in a pectic substance apparently plays an important role in the capacity of a given pectin to form a good jelly. A fully esterified pectinic acid (not found in nature) should contain, theoretically, 16% methoxyl groups. Naturally obtained pectins contain from 9.5 to 11% methoxyl groups and, if esterified to the extent of about 8% ester, will give pectinic acids best suitable for preparing jellies.

Because the degree of methylation is not the only measure of the jellifying property of the pectins and because such other qualities as the size of the molecule, etc., are rather complicated to be determined in industrial usage, commectial pectins are evaluated by "*pectin grades*" and are expressed as the number of parts of sugar that one part of pectin will gel to an accepted firmness under standard conditions. These standards are accepted to be a pH of between 3.2 to 3.5, 65 to 70% sugar and pectin in the limits of 0.2 to 1.5%. In commerce, pectins of 100 grade up to 500 grade are marketed. They also differ in their "setting time"; a "*rapid* set" begins at about 85° C while a "*slow* set" forms a jelly below 55° C. The rapid set is used in the manufacture of preserves in order to prevent whole fruit or chunks settling to the bottom or rising to the top of the jar instead of being evenly distributed throughout the jam.

Commercial pectins are characterized according to their *jellying power* (grade), *degree of methoxylation* (high methoxyl or low methoxyl pectins) and the *rate of solidification* of the jellies (rapid, medium, slow set pectins). Other factors being equal the jellying power increases with increasing molecular weight. The degree of methoxylation actually determines the mechanism of gel formation. The rate and temperature of setting is also governed by the extent of esterification, rapid set pectins having a higher degree of methoxylation. At equal degrees of esterification, pectins with a higher degree of polymerization require shorter setting times (Doesburg and Grevers).

As all lyophilic high polymers, pectins impart considerably higher viscosities to their solutions. Incidentally, the viscosity of pectin solutions is strongly affected by their pH. At higher pH the free carboxyl groups undergo dissociation and form negatively charged centers which repel each other. As a result of this, the molecules assume an unfolded, straight, rigid configuration, thus increasing viscosity. Obviously, it can be expected that the effect of pH on viscosity would be stronger in solutions of low methoxyl pectin.

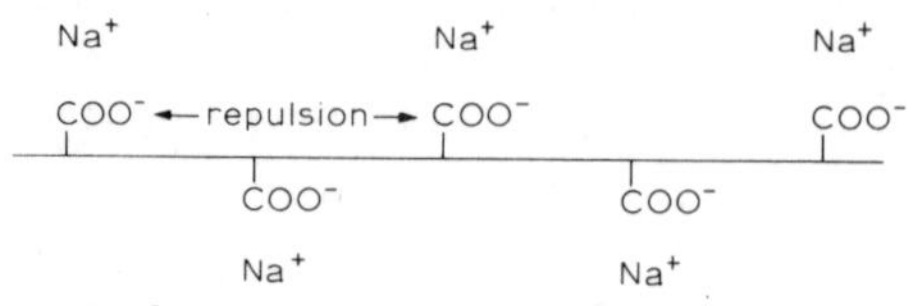

As long as pectin molecules remain as distinct chains, the solution may be quite viscous but never solidifies into a rigid gel. Solidification requires the formation of a tridimensional structure of chains, immobilizing large portions of the liquid within the network. Pectin molecules may form such a structure by means of hydrogen bonds between chains. However, in a neutral aqueous solution of pectin, two factors prevent intermolecular hydrogen bonding. First, pectin is strongly hydrated, i.e. the conditions are more favorable for the existence of pectin–water bonds than pectin–pectin bonds. Second, as partially ionized polyelectrolytes, the molecules of pectin repel each other. In order to counteract these two factors, the pectin must be dehydrated and its dissociation depressed. In jams and jellies, sugar (and sometimes polyols such as sorbitol) is used as a suitable dehydrating agent. Depression of ionic dissociation is achieved by lowering the pH.

In the case of low methoxyl pectins the tridimensional structure can be formed by means of Ca^{++} bridges between free carboxyl groups. This type of structure, sometimes called the "*calcium pectate gel*", is considerably more rigid than the pectin–sugar–acid gel. No sugar or acid is needed for the formation of the calcium pectate type gel, hence the use of low-methoxyl pectins (in conjunction with a soluble calcium salt) in the manufacture of low-sugar preserves for dietetic purposes.

To summarize, the special gel systems obtained with pectins of various degrees of methoxylation are as follows:

(a) Completely methoxylated polygalacturonic acid will form gels with sugar alone. Here the gelation is caused solely by the dehydrating effect of the sugar. No acid is needed since no dissociated carboxyl groups are present.

(b) Rapid set pectins are pectins with a degree of methoxylation of 70% and higher. They will form gels on addition of acid and sugar at a pH optimum of 3.0 to 3.4 at relatively high temperatures. The strength of such gels is largely dependent upon the molecular weight and is not influenced by the degree of methoxylation. The higher the molecular weight the stronger is the gel.

(c) Slow set pectins are those with a degree of methoxylation of 50–70%. They will form gels on addition of sugar and more acid at a pH optimum of 2.8–3.2 and at lower temperatures. The amount of acid required is approximately proportional to the number of free carboxyl groups.

(d) Low-methoxyl pectins are pectins with a degree of methoxylation under 50%. They do not form jellies with sugar within acceptable limits of acidity but will gel in the presence of calcium ions and other polyvalent cations. The amount of pectin required for the formation of such gels decreases with the degree of methoxylation. The strengths of such ionic-bonded gels are strongly dependent upon the degree of esterification and are only little affected by the molecular weight of the pectins.

9-5. CLARIFICATION OF CLOUDY JUICES

Fruit juice manufacturers are often confronted with two conflicting problems. On one hand, it is frequently important to clarify, i.e. to remove suspended particles from products such as wines and clear apple juice. On the other hand, it is sometimes desired to retain the cloudy appearance of the product and to avoid separation (citrus juices, tomato products, some soft drinks). Both problems seem to be closely related to pectic substances and pectolytic enzymes.

Cloudy juices consist of a suspension of very fine particles (often less than 1 micron in any direction) in a clear medium (serum). The chemical composition and physical properties of both cloud and serum may vary considerably from one fruit to another and each system must be studied separately.

One of the stabilizing factors in cloudy juices may be the electrostatic charge of the suspended particles. *Orange juice cloud* is negatively charged. The site of the negative charge is believed to be dissociated free carboxyl groups of the pectin present on the surface of the particles (Mizrahi and Berk). Deuel and his co-workers took advantage of the fact that such colloids with a negative charge will create insoluble complexes with other soluble polyelectrolytes of opposite charge and will then precipitate easily. Accordingly, they suggested the use of polyethylenamine for the clarification of cloudy wines and juices.

$$\text{Pectinate}^- + \text{Polyethylenamine}^+ \rightarrow \text{Complex} \downarrow$$

The precipitate thus formed will soon settle, carrying with it all suspended matters as well as tannins, which may be adsorbed upon large molecules of this kind.

Another method for attaining similar effects consists of treating cloudy liquids with commercial preparations of pectolytic enzymes, such as "Pectinol". These preparations are usually obtained from molds and contain PE and endo-PG.

When, however, the opposite is required, i.e. a permanent cloud stability, it is then necessary to prevent, as far as possible, the de-esterification of the pectin. Citrus juices must be heat-treated as soon as possible after juice extraction, to inactivate PE and prevent clarification. In the case of tomato juice, the pectolytic enzymes act so quickly that breakdown of the colloidal structure occurs unless the fruit is heated rapidly before or right at the moment of crushing (*hot break*).

The mechanism of clarification by pectolytic enzymes is not known. It is interesting to note that PE alone is responsible for the clarification of citrus juice since this material does not exhibit any polygalacturonase activity. Older theories suggest that soluble pectin is necessary to keep the cloud in suspension. Baker and Bruemmer reject this view, on the basis of experimental evidence. According to these authors, orange cloud will stay in stable suspension in synthetic media devoid of soluble pectin or in depectinized orange serum, even in the presence of PE activity. Thus soluble pectin would be a cause of instability rather than a stabilizing factor, as far as cloud is concerned. The mechanism of clarification would then be: (a) de-esterification of soluble pectin by PE action; (b) formation of insoluble pectates in the presence of calcium and magnesium ions; (c) precipitation of these complexes carrying down all the suspended material.

It should be pointed out that this interesting theory is in contradiction with experimental findings by other investigators. Yufera et al. found that considerable demethoxylation of the insoluble pectin contained in the suspended pulp occurs in parallel with clarification of orange concentrate. They conclude that clarification is due to a modification of the surface properties of the suspended particles rather than changes in the continuous medium.

Still a different mechanism is suggested by Yamasaki and co-workers, for apple juice. According to these investigators suspended particles in cloudy apple juice consist of a protein rich core (positively charged at the acidic pH of the juice) surrounded by a layer of negatively charged pectin. The serum itself contains negatively charged pectin in colloidal solution. Under the action of pectolytic enzymes (PE and PG), the pectin layer on the surface of the particle is destroyed and the protein core with its positive charge is exposed. Now coagulation occurs, as a result of electrostatic neutralization between the two colloids of opposite charge.

In summary, pectolytic enzymes are responsible for clarification in cloudy juices. In the case of concentrated citrus juice the formation of calcium pectate does not cause clarification but actual gelation of the product. The mechanism of action of the enzymes in clarification remains to be elucidated.

9-6. USE OF PECTINS IN FOOD

As already mentioned, the main use of pectin is in the manufacture of jams, jellies, marmalades and fruit preserves of all kinds, especially when fruits of naturally poor pectin content are used. In the natural juice products, pectin can be used for strengthening the permanent cloud stability, for raising the viscosity of tomato products and for similar purposes.

When canning fruits which are somewhat over-ripe, their texture may be too soft because of the degradation of the natural pectic substances having advanced too far. In such cases, use is often made of calcium salts, which will create Ca-pectate in the cells of the fruit and thereby strengthen the texture.

Low methoxyl pectin is now extensively used in the manufacture of canned fruits such as grapefruit segments, cherries, peaches. When added to the syrup used for canning, low methoxyl pectin is found to increase firmness and drained weight of the fruit.

Second only to its ability to form gels is the very important property of pectin to act as an emulsifier. Pectin makes very good emulsions with edible oils for the production of mayonnaise, ice cream and also with essential oils when used in the manufacture of various flavors.

Good use is made of pectins in medicine as cleaners for the intestines, for treating wounds and in blood transfusions for raising the blood volume. Some use of pectin is also made in industry for steel hardening, etc. For a comprehensive description of such uses, together with a discussion of problems relating to pectic substances and their enzymes, the reader is referred to the classical treatise on this subject by Kertesz.

According to Doesburg and De Vos, milk to which high methoxyl pectin is added will not undergo curdling when acidified to a pH as low as 3.0, even if the mixture is pasteurized after acidification. This may serve as a method for the preparation of stable milk-fruit juice mixtures.

Another interesting recent development is the possibility of using pectin gels as protective films for coating foods. One instance of this type of use is the preservation of fresh grapefruit halves by coating them with a gel containing low methoxyl pectin and calcium chloride.

9-7. INDUSTRIAL MANUFACTURE OF PECTINS

Commercial pectins are manufactured in two main forms: "liquid pectins", which present more or less concentrated solutions of pectin extracted from waste plant materials, such as apple pomace or citrus peels, and "dry pectin powders". In commercial practice the resulting products are not pure substances; their degree of purity, or rather their evaluation, depends largely on the methods of manufacture, the molecular size of the pectic substances, the degree of esterification and the amount of accompanying ballast material present in the final product.

The amount of pectin in the raw material greatly varies for different varieties. Thus, for instance, among citrus fruits the most pectin-rich varieties are grapefruit, lemon and pomelo. Secondly, it has been shown that unripe fruit is richer in pectin than fully ripe apples or citrus because, as the season advances, the quantity of effective pectin in the fruit diminishes.

The manufacturing procedures comprise the following main steps:

(1) Preparation of material.
(2) Removal of ballast.
(3) Acid hydrolysis of protopectin and dissolution of pectin.
(4) Precipitation.
(5) Purification and drying.

Detailed methods of pectin manufacture have been described by Braverman. These are mainly based on two procedures: the precipitation of pectins by alcohols or acetone, or by making use of the fact that some mineral salts, such as aluminium hydroxide, carry in their colloidal state an electric charge with opposite sign to the negatively charged pectinic acids and are, therefore, able to coagulate the pectin from its aqueous solutions. In this method ammonia is added to the filtered aqueous extract of pectin to bring its pH to 4 and then sulfate of alumina added:

$$Al_2(SO_4)_3 + NH_4OH \rightarrow Al(OH)_3 + (NH_4)_2SO_4$$

When the precipitation of pectin has been completed the free liquid is drained off in false-bottom tanks and, aided by agitators, the precipitated gel is collected, pressed between cloth and treated with acidified alcohol–water solutions in order to remove excess $Al(OH)_3$. Repeated washings with acidified isobutyl alcohol and finally with water-free alcohol give a more or less pure gel which is pressed, dried and ground, while the alcohol is regenerated by distillation. The grade of pectin obtained in this process is about 300.

A number of processes using ion exchange and others have been patented in many countries.

9-8. PECTINS IN ALCOHOLIC FERMENTATION

When ethanol is produced from waste plant materials containing pectins, care must be taken that the pectins are not de-esterified before or during fermentation, otherwise there is a danger of *methanol* being present in the final product. Such instances may occur when the plant material is limed before fermentation, as the case may be during the extraction of citrus-peel juice, for instance. Similar difficulties may arise during the manufacture of fermented apple cider. If the fruit or other plant material becomes contaminated with mold (a rich source of PE), the amount of methanol formed may be fairly high.

9-9. OTHER PLANT GUMS

Strictly speaking, *gums* are substances exuded by plants as a response to tissue injury. In a more general context, the words "plant gums" indicate a group of water dispersible complex polysaccharides of plant origin, used extensively in food and other industries as thickeners, binders and stabilizers. "*Plant hydrocolloids*" is probably a better name for this group.

Plant hydrocolloids are generally heteropolysaccharides. The heterogeneous structure of their chain inhibits the organization of crystallites and makes these substances readily soluble in hot water. The high molecular weight and affinity to water set the basis for the most important property of the gums, i.e. that of forming very viscous solutions or firm gels even at low concentration. By this definition, pectic substances and the synthetic carboxymethyl cellulose (Section 8-2) are also gums.

Sodium alginate is the sodium salt of *alginic acid* extracted from algae. It is widely used as a stabilizer in frozen desserts and ice creams; it has also exceptional foam stabilization properties in beverages. Alginic acid is not a heteropolysaccharide but a mixture of two practically homogeneous polymers: poly-O-mannuronic acid and poly-L-guluronic acid.

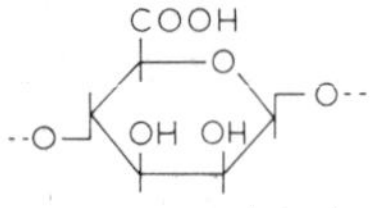

D-mannuronic acid residue

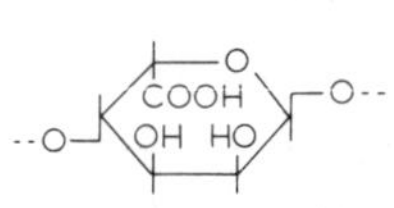

L-guluronic acid residue

Alginic acid itself is insoluble in water, therefore sodium alginate cannot be used in acid foods. In this case a propylene glycol derivative is used.

Carrageenan and agar are both extracted from sea plants. Both substances are sulfated galactans. *Carrageenan* or Irish moss extract, is a mixture of two fractions, κ-carrageenan and λ-carrageenan. The first consists of residues D-galactose and 3-6- anhydro-D-galactose. Sulfuric acid is bound through ester linkage at positions 4, 2 or 6. λ-Carrageenan is a highly sulfated complex galactan. Due to the presence of sulfuric acid groups, carrageenan is strongly acid and forms a negatively charged polymer, which reacts with positively charged protein molecules to give viscous solutions or gels.

Agar is extensively used for the solidification of nutrient media in bacteriology. It is sometimes utilized in food jellies (flans), instead of gelatin. It is a complex polymer containing mainly D-galactose but also a sugar with L configuration: 3,6-anhydro-L-galactose. Agar also contains much less sulfate than carrageenan.

Karaya gum is the dried exudation of the *Sterculia urens* tree grown in India. While its structure has not yet been elucidated, it is considered to be an acetylated polysaccharide of high molecular weight. Of all the hydrocolloids, Karaya gum is the least soluble, but, when finely ground, it readily absorbs· water. Karaya is used as a stabilizer in French dressings, in ice popicles and sherbets, where it prevents syneresis as well as the formation of large ice crystals. In order to prevent water separation and to increase the ease of spreading, Karaya is also used in the manufacture of cheese spreads. In addition, this gum imparts a smooth appearance to ground meat products which require an efficient water-holding capacity on the other hand.

Gum tragacanth is a water-soluble exudate obtained from *Astragalus gummifer* and forms a thick viscous liquid. Tragacanth gum has been shown to have a complex structure containing D-galacturonic acid as a main chain to which several pentoses (L-fucopyranose, D-xylopyranose, D-galactose and L-arabinose) are linked in glycosidic bonds. Its ability to swell in water to form gels of high water content makes it very useful as a stabilizer in puddings, salad dressings, mayonnaise and ice cream manufacture.

Both *locust bean gum* and *guar gum* are produced from the plant family Leguminose: the former from the seed endosperms of the carob tree, *Ceratonia siliqua* L., which grows in Mediterranean countries, and the latter from the *guar*, a legume resembling the soybean plant. Both gums are galactomannans, i.e. long $1 \rightarrow 4$ chains of polymannans to which single galactose units are attached through $1 \rightarrow 6$ glycosidic bonds:

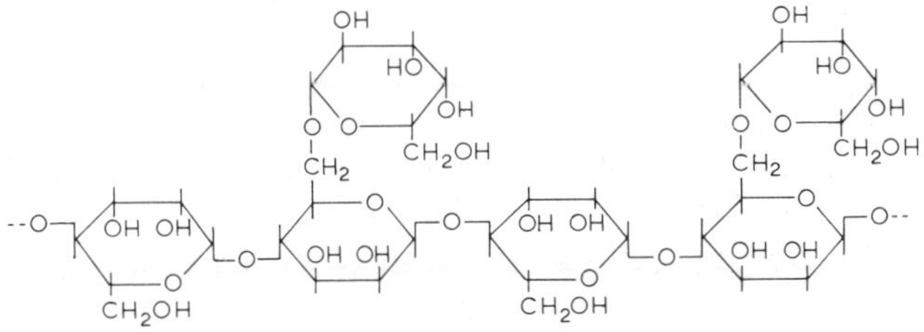

portion of guar gum gallacto mannan

These gums swell in cold solutions and require no heat to complete their hydration. Because of their strong hydrophilic character they are excellent additives in salad dressings, ice cream mixes and bakery products. They are also used in the paper and textile industries.

9-10. NUTRITIONAL AND TOXICOLOGICAL ASPECTS

The beneficial effects of pectin (detoxification, intestinal regulation, etc.), have been mentioned in Section 9-6. This is probably the basis for the use of pectic substances in pharmacy and for the well known saying: "*an apple a day, etc. . . . .*" More recently, it has been reported that ingestion of large quantities of pectic substances results in lower blood cholesterol levels. There are still differences of opinion on the validity of these observations as on the possible mechanism of action.

Plant gums are practically indigestible and, therefore, they have been long considered as physiologically "inert" and safe for human consumption. With the increasing usage of gums as stabilizers and thickening agents in modern food technology, this view is now open for more critical examination. It has been found, for example, that certain modified carrageenans may cause severe lesions in the intestinal wall of experimental animals. The relevance of these findings to the practical use of gums as food additives is not clear and many of these materials continue to be "generally recognized as safe".

SELECTED BIBLIOGRAPHY TO CHAPTER 9

American Chemical Society, Natural Plant Hydrocolloids, Advances in Chemistry Science No. 11, Washington D.C., 1954.

Baker, R.A. and Bruemmer, J.H., Could Stability in the Absence of Various Orange Juice Soluble Components, Florida State Hort. Soc., **82** (1969), 215-220.

Braverman, J.B.S., Citrus Products—Chemical Composition and Technology, Interscience, New York, 1949.

Doesburg, J.J., Pectic Substances in Fresh and Preserved Fruits and Vegetables, I.B.V.T., Communication No. 25, Wageningen, 1965.

Jansen, E.F., Jang, R. and Bonner, J., Orange Pectinesterase Binding and Activity, J. Amer. Chem. Soc., **68** (1946), 1475.

Kertesz, Z.I., The Pectic Substances, Interscience, New York, 1951.

Whistler, R.L. and Bemiller, J.N. (Eds.), Industrial Gums, Academic Press, New York, 1959.

Worth, H.G.J., The Chemistry and Biochemistry of Pectic Substances, Chem. Rev., **67** (1967), 465-473.
Yamashaki, M., Yasui, T. and Arima, K., Pectic Enzymes in the Clarification of Apple Juice, Agr. Biol. Chem., **31** (1964), 552-560.
Yufera, E.P., Mosse, J.K. and Royo-Iranzo, J., Gelification in Concentrated Orange Juice, Inst. Intern. Congress Food Science Techn., **2** (1965), 337-343.

CHAPTER 10

NON-ENZYMIC BROWNING

The Pandora Box

10-1. INTRODUCTION

The formation of dark colored pigments in foods during processing and storage is a very common phenomenon. The subject is of primordial interest because it involves not only the color and appearance of the food, but its flavor and nutritive value as well. In a number of cases such as the manufacture of malt and maple syrup or roasting of coffee, cocoa and nuts, or toasting of breakfast cereals, or baking of bread (crust), etc., the production of the dark color and accompanying changes in flavor are desirable. However, as a rule, browning of food products is a distinct sign of deterioration. Browning, more than any other alteration, is the reason for the commercial "death" of many foods, the most important factor limiting their useful shelf life.

Although the end results are practically the same, reactions leading to browning discoloration are extremely varied and complex. Some are catalyzed by enzymes and involve oxidative reactions in which phenolic compounds participate. This class, known as "enzymic browning", will be described in a later chapter. For the present, we shall address our attention to non-enzymic browning reactions. Examples of this type of discoloration are numerous; they appear in dry egg and milk powders, in dehydrated vegetables, in canned and dried fruit, in fruit juices and concentrates, in starch, corn syrup, protein hydrolysates, dehydrated meat and fish, etc.

Although many aspects of these phenomena have not been thoroughly elucidated, especially as far as the later stages of pigment formation are concerned, it is assumed that three distinct mechanisms are involved in non-enzymic browning:

(a) The Maillard reaction involving interaction between sugars and amino acids (free or combined in peptides and proteins).

(b) Reactions involving ascorbic acid.

(c) Caramelization of sugars with or without the catalytic action of acids.

There is no doubt that in foods which are complex systems, combination and interactions of all three paths often occur. For this reason, investigation of browning reactions in actual foods is usually difficult. Therefore, most of our present knowledge on the reactions leading to browning is based on experiments on model systems, consisting of a small number of pure substances.

10-2. GENERAL DESCRIPTION

The most significant components of foods, as far as participation in non-enzymic browning is concerned, are low molecular weight carbohydrates and their derivatives (sugars, sugar acids, ascorbic acid). Free amino acids and the free amino groups of proteins and peptides participate in reactions leading to the formation of brown pigments; however, browning may also occur in the absence of nitrogenous substances.

The products of the first reactions leading to browning are colorless. At some stage, products with characteristic absorption spectra in the U.V. range are formed. As a rule, the progress of the reaction is characterized by the formation of increasingly reducing substances. Unsaturated bonds and true carbonylic groups are formed and their concentration increases. Water is produced as a result of successive dehydration steps involving the carbohydrate participants. With regard to molecular weight of the reactants, degradation and fission processes occur at some stage. CO_2 is evolved. Volatile components formed as a result of these fission reactions are responsible for the production of characteristic flavors, which accompanies browning reactions. The final steps of the process, however, involve polymerization reactions as a result of which the dark pigments are produced. As polymerization proceeds, the pigments become darker and less soluble in water.

The initial stages of the browning process have been elucidated in a large number of model systems and some foods. However, even when one starts with a relatively simple model system, such as a solution of glycine and glucose, the number of intermediate products and possible reaction pathways soon reaches bewildering proportions.

10-3. THE MAILLARD REACTION

The French chemist Maillard was the first to study the condensation of sugars with amino acids. He reported in 1912 that when mixtures of amino acids and sugars are heated, brown substances are formed. Since then, the Maillard reaction has been considered as the major cause of non-enzymic browning in foods and a vast amount of experimental data has been suggested as proof of this connection.

The course of non-enzymic browning as a consequence of the Maillard reaction may be divided into the following steps:

(a) Sugar–amino condensation.
(b) Rearrangement of condensation products.
(c) Dehydration of rearrangement products.
(d) Further (or parallel) fission and degradation.
(c) Polymerization into pigments.

(a) *Condensation of sugars with amino compounds*

It will be recalled (Section 6-4-4) that sugars react with primary and secondary amines to form *glycosylamines.* Similar condensation reactions occur with free amino acids and with free amino groups of peptides and proteins. This is believed to be the first step of the Maillard reaction. This step is represented below for the condensation of glucose with glycine.

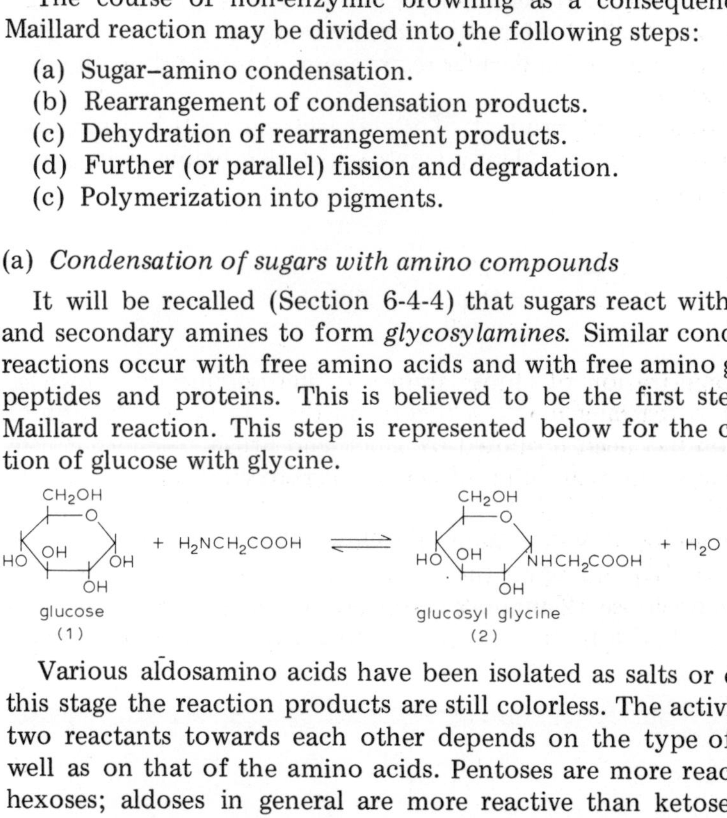

CH_2OH, HO, OH, OH, OH + H_2NCH_2COOH $\rightleftharpoons$ CH_2OH, HO, OH, $NHCH_2COOH$, OH + H_2O

glucose (1) — glucosyl glycine (2)

Various aldosamino acids have been isolated as salts or esters. At this stage the reaction products are still colorless. The activity of the two reactants towards each other depends on the type of sugar as well as on that of the amino acids. Pentoses are more reactive than hexoses; aldoses in general are more reactive than ketoses; monosaccharides are more reactive than disaccharides. Obviously, for the condensation to occur, the sugar must have a free glycosidic OH, which explains the relative inertness of sucrose and other non-reducing sugars. Uronic acids and phosphate esters condense with amines more readily than does the corresponding sugar. As to amino acids, strongly basic ones have been found to be more reactive.

The condensation reaction itself is reversible and glycosylamino compounds are easily hydrolyzed by dilute acids. As the NH_2 group of the amine, amino acid or protein is blocked by the condensation reaction, one of the effects of the process is a drop in pH. As may be expected, the condensation rate is enhanced by high pH.

(b) *Rearrangement of the condensation products*

When aldoses are reacted with amino compounds, the reaction mixture is soon found to contain ketose derivatives. Inversely, aldose derivatives are formed when the starting material is a ketose. It is evident that a stage of molecular rearrangement must follow the step of condensation.

The rearrangement product of a glucosylamine (3) is a *fructose-amine* (4) (or 1-amino-1 deoxy fructose), shown below in its stable pyranoside form:

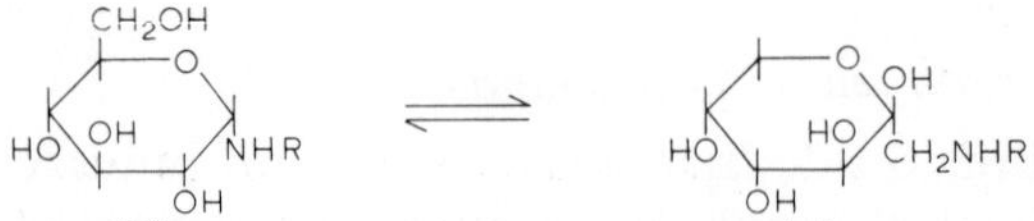

CH_2OH O HO OH NHR OH (3) ⇌ O OH HO HO CH_2NHR OH (4)

The isomerization of aldosylamines to ketosamines is known as the Amadori rearrangement. A theory on the possible mechanism of the reaction has been proposed by Isbell and Frush. The mechanism also explains the catalytic effect of carboxylic acids in molar quantities.

Isomerization of ketosylamines follows the so-called Heyns re-arrangement. Depending on the conditions, the reaction product can be a 2-aminoaldose (2-amino-2-deoxyaldose) or a ketosamine. The rearrangement of a fructosylamine (5) to a 2-amino glucose is shown below.

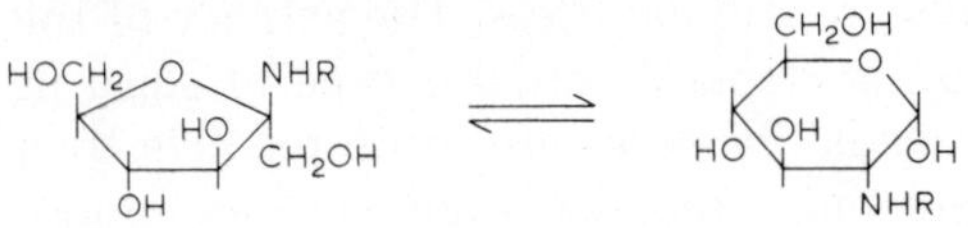

$HOCH_2$ O NHR HO CH_2OH OH (5) ⇌ CH_2OH O HO OH OH NHR (6)

The formation of ketosamines is explained by Heyns as a secondary reaction involving condensation of an additional amine group with the glycosidic OH now available, then an Amadori rearrangement and finally, hydrolytic cleavage of the amino group on the 2 position.

The Amadori or Heyns isomerizations are reversible and the reaction products, aldosamines or ketosamines, are still colorless.

(c) *Dehydration of rearrangement products*

What exactly follows the rearrangement step is not quite clear. A large number of intermediate decomposition products have been identified both in model systems and in foods. It is clear, however, that increasingly unsaturated, highly reactive intermediates are formed. One of the products frequently found is *furfural* or its derivatives. It is known that pure aldosamines and ketosamines undergo spontaneous decomposition under certain conditions, but the rate of reaction is too slow to be significant in actual browning. Reynolds postulated that the rearrangement products must be first transformed into other less stable intermediates. One possibility is the formation of *diketosamines*, such as difructosamine (7), by condensation of the ketosamine with an additional molecule of sugar, followed by rearrangement.

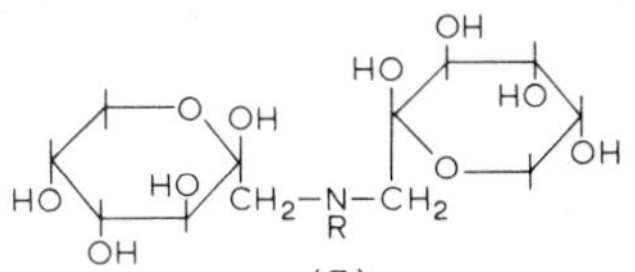

OH HO HO O OH OH O HO HO CH_2–N–CH_2 R OH

(7)

Diketosamines are less stable. When heated in aqueous solution they yield monoketosamine and a number of decomposition products, with marked carbonylic and unsaturated properties. Thus the formation of diketosamines may be viewed as the catalytic activation of a sugar molecule by a molecule of ketosamine.

Although at a slower rate, ketosamines also undergo the same type of degradation as the diketosamines. The carbonylic unsaturated intermediates are osuloses and the final products of this stage are furfural derivatives, mainly *hydroxy methyl furfural* (HMF) when the starting material is a hexose. Several routes have been suggested for the formation of osuloses and furfural, starting with fructosamines. One of the possible courses is shown below.

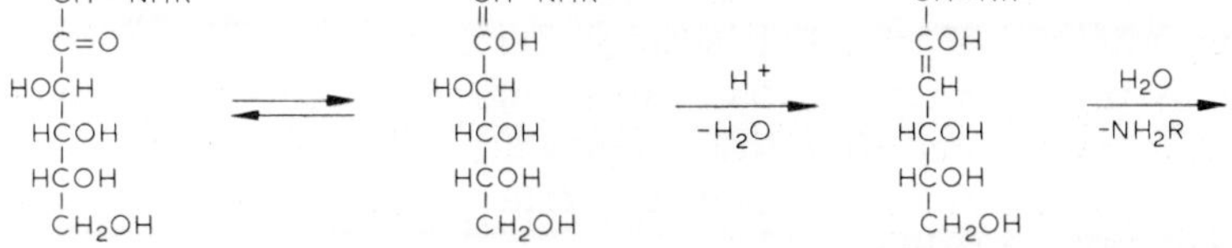

(8)		(9)		(10)	
CH –NHR		CH–NHR		CH=NR	
C=O		‖ COH		COH	
HOCH	⇌	HOCH	$\xrightarrow[-H_2O]{H^+}$	‖ CH	$\xrightarrow[-NH_2R]{H_2O}$
HCOH		HCOH		HCOH	
HCOH		HCOH		HCOH	
CH_2OH		CH_2OH		CH_2OH	

(11) CHO–COH=CH–HCOH–HCOH–CH_2OH $\xrightarrow{-H_2O}$ (12) CHO–C=O–CH=CH–HCOH–CH_2OH $\xrightarrow{-H_2O}$ (13) CHO–C=CH–CH=C(–O–)–CH_2OH

In the first place, fructosamine (8) isomerizes to its enolic form (9). The reaction is reversible. Now the OH group in position 3 is eliminated and a Schiff base (10) is formed. Hydrolysis of the latter yields a *3-deoxy osulose* (11), shown in its enolic form. Elimination of the hydroxyl from C-4 by dehydration gives an unsaturated *osulose* (12), which is transformed to HMF (13) by removal of an additional molecule of water. The net result of this course may be summarized as the liberation of the amino moiety and gradual dehydration of the sugar derivative. The elimination of the nitrogenous group is not essential. It may remain in the molecule, in which case the final product is not HMF but a Schiff base of HMF (14).

CH=NR, HC, HC, O, CH_2OH

(14)

At this point, it is useful to compare the course of actual browning phenomena in foods with the reactions described so far. When a food which is known to be susceptible to non-enzymic browning is stored, significant changes occur in the amino acid content already after a relatively short time. Some amino acids disappear at a higher rate than others. The rate of amino acid loss, on a molar basis, is usually lower than that of sugar loss. This may be proof of the diketosamine route. HMF often accumulates, soon before the appearance of dark pigmentation. Spectrophotometric determination of HMF at this stage is often used as a standard technique for the prediction of browning rate in accelerated storage tests. Osuloses have been detected as transient intermediates in a number of browning foods. Darkening occurs soon after the appearance of HMF and a strong correlation usually exists between the rate of HMF formation and the intensity of browning.

So far, the mechanism of the initial reactions was described in relation to model systems consisting of sugars and free amino acids. Reactions involving proteins are essentially similar. Aldoses and ketoses condense with the free amino group of proteins (ϵ-amino groups of lysine, terminal amino groups). This has been demonstrated in model systems consisting of carbohydrates and pure proteins of known amino acid sequence. For instance, working with insulin–glucose mixtures Schwarts and Lea demonstrated that the terminal glycyl and phenylanyl amino groups and lysine ϵ-amino groups combine with the aldose.

(d) *Degradation and fission reactions*

Carbon dioxide production in the course of amine–sugar interaction was already observed by Maillard. The formation of low-molecular weight intermediates such as pyruvaldehyde and diacetyl in sugar–amino acid mixtures was reported by Speck and confirmed by many others. Obviously, these compounds must originate from the fragmentation of the carbohydrate moiety, or the amino acid or both.

In the presence of certain dicarbonylic compounds, amino acids are known to undergo decarboxylation as follows:

$$R_1-\overset{O}{\overset{\|}{C}}-\overset{O}{\overset{\|}{C}}=R_2 + R_3 - CH(NH_2)COOH \longrightarrow R_1-C(NH_2)=C(OH)-R_2 + R_3CHO + CO_2$$

This reaction, known as the Strecker degradation, results in the simultaneous deamination and decarboxylation of the amino acid into an aldehyde. The importance of Strecker degradation in non-enzymic browning is now widely recognized. By using isotope tracers, it has been demonstrated that 90–100% of the expelled CO_2 in Maillard reaction originates from the amino acid moiety and not from the sugar residue. The migration of NH_2 from the amino acid to the dicarbonyl is regarded by Hodge as a possible mechanism by which nitrogen is incorporated into the brown pigments. As to the nature of the dicarbonyl or potential dicarbonyl partner in the Strecker degradation, osuloses, Amadori rearrangement products and reductones (see further), may well fulfil this function.

With respect to the fission of carbohydrate material, Hodge proposes *dealdolization*, i.e. the reverse of aldol condensation as the most probable mechanism. Aldol condensation is the self-condensation of two aldehyde molecules as follows:

$$R_1CH_2-C(=O)H \quad + \quad O=CR_2H \rightleftharpoons R_1-CH(CHO)-CH(OH)-R_2$$

carbonyl I　　　carbonyl II　　　aldol

The addition product usually undergoes dehydration into the more stable unsaturated aldehyde form:

$$R_1-CH(CHO)-HCOH-R_2 \xrightarrow{-H_2O} R_1-C(CHO)=CH-R_2$$

Aldol condensation is catalyzed by amines, which will also catalyze the reverse reaction, i.e. dealdolization of carbohydrates, especially the products of Amadori rearrangement.

(e) *Polymerization into brown pigments*

We have seen how a relatively simple model system consisting of a sugar and an amino acid soon becomes a complex mixture of a multitude of highly reactive compounds: furfural and derivatives, deoxy and unsaturated osuloses, aldehydes from Strecker degradation, aldehydes and ketones from sugar fission, etc. The fact that most of these compounds themselves are known to undergo rapid browning in the presence of amino acids may be taken as proof of their role as intermediates in the process of pigment production.

The final stages of the browning reaction apparently involve random polymerization of these carbonylic intermediates, with or without the participation of amino acids. Polymerization of the carbonyl compounds may follow successive aldol condensation reactions (see above), catalyzed by the presence of amino acids. Co-polymerization of carbonyl intermediates with available amino groups may be the mechanism by which proteins become attached to the brown pigments.

The final products of the reaction, the brown *melanoidins*, are complex high molecular weight pigments of unknown structure. At the early stages of polymerization, the pigments are water soluble. They have featureless absorption spectra in the visible range; their absorbance increases with decreasing wavelength, in a continuous, peakless fashion. Log-log linear relationship between wavelength and absorption is often observed. Infrared spectra and chemical tests indicate the presence of unsaturated bonds, heterocyclic structures, some intact amino acid residues. Purification of the pigment mixtures and their separation into more or less homogenous fractions are difficult operations. Considerable progress in this field has been reported after the development of gel filtration techniques. The elemental analysis of the pigments reveal various empirical formulae, such as $C_{18}H_{27}O_{12}N$ (from maple syrup), $C_{19}H_{30}O_{16}N_3$ (from milk powder), $C_{17}H_{19}O_9N$ (from dried apricots), etc. These formulae provide no information on the structure of the pigments. Repetitive patterns have been sometimes reported but random polymerization is apparently the more general case. In systems containing proteins, the latter is bound to the dark pigment, usually through the ϵ-amino of lysine. The protein moiety may be removed from the pigment by hydrolysis, in which case a *limit-peptide pigment* complex (LLP) is obtainedd. Such a peptide pigment from a casein–glucose system has been studied by Clark and Tannenbaum. The nutritional and organoleptic consequences of the participation of proteins in Maillard reaction will be discussed later.

Browning via reductones

The sequence of reactions described above involved 1,2-enolization of the Amadori rearrangement products. An alternative route to brown pigments, based on 2,3-enolization, has been suggested. According to Hodge, this route is the predominant mechanism of sugar dehydration in low-moisture systems. The intermediates of this mechanism are *reductones*, i.e. compounds having the following structure:

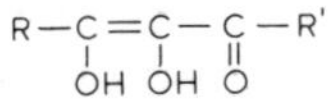

$$\mathrm{R-\underset{\displaystyle OH}{\underset{|}{C}}=\underset{\displaystyle OH}{\underset{|}{C}}-\underset{\displaystyle O}{\underset{\|}{C}}-R'}$$

The sequence of reactions leading to the production of reductones may be summarized as follows, starting with a ketosamine (15) from Amadori rearrangement:

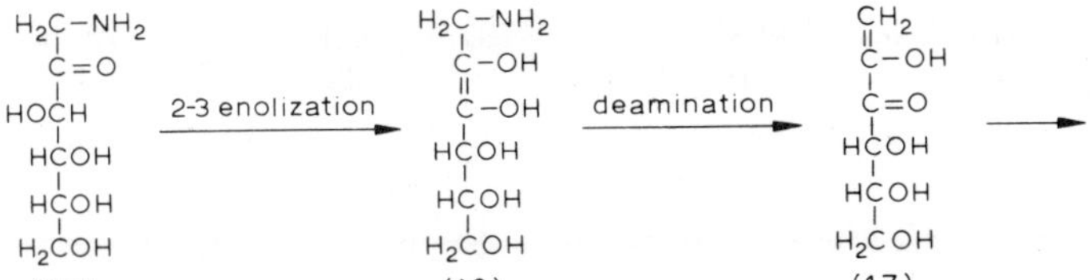

$$\begin{array}{c} H_2C{-}NH_2 \\ | \\ C{=}O \\ | \\ HOCH \\ | \\ HCOH \\ | \\ HCOH \\ | \\ H_2COH \\ (15) \end{array} \xrightarrow{\text{2-3 enolization}} \begin{array}{c} H_2C{-}NH_2 \\ | \\ C{-}OH \\ \| \\ C{-}OH \\ | \\ HCOH \\ | \\ HCOH \\ | \\ H_2COH \\ (16) \end{array} \xrightarrow{\text{deamination}} \begin{array}{c} CH_2 \\ \| \\ C{-}OH \\ | \\ C{=}O \\ | \\ HCOH \\ | \\ HCOH \\ | \\ H_2COH \\ (17) \end{array} \longrightarrow$$

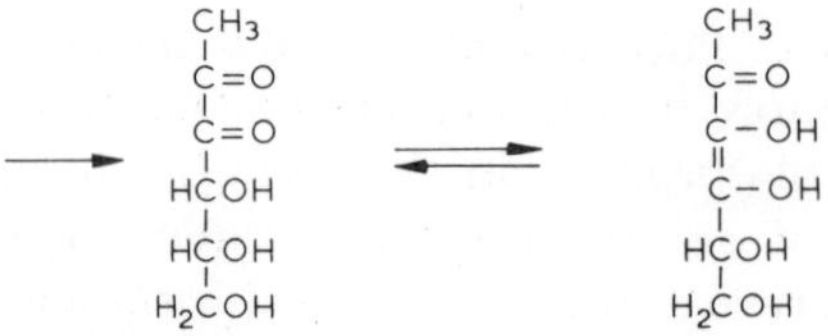

$$
\longrightarrow
\begin{array}{c}
CH_3 \\
| \\
C{=}O \\
| \\
C{=}O \\
| \\
HCOH \\
| \\
HCOH \\
| \\
H_2COH \\
(18)
\end{array}
\quad \rightleftharpoons \quad
\begin{array}{c}
CH_3 \\
| \\
C{=}O \\
| \\
C{-}OH \\
\| \\
C{-}OH \\
| \\
HCOH \\
| \\
H_2COH \\
(19)
\end{array}
$$

Reductones such as (19) presumably undergo dehydration, condense with amines and finally polymerize to melanoidins. They may also break down to produce smaller molecules such as diacetyl, acetic acid, pyruvaldehyde, or they may form more stable reductones, containing conjugated double bonds. Such reductones have been isolated from mixtures of several sugars with secondary amines.

It will be noted that compound (19) contains a C-methyl group which was not encountered as an intermediate of the 1-2 enolization route. The presence of C-methyl groups in the brown pigments from some model systems has been suggested as a proof of the validity of the 2-3 enolization mechanism.

10-4. BROWNING OF ASCORBIC ACID

Ascorbic acid is responsible for a good part of browning phenomena in fruit juices and concentrates. Of particular importance is the role of ascorbic acid in the discoloration of citrus products. Model solutions containing ascorbic acid in the presence of amino acids darken more rapidly than do sugar–amino acid systems under the same conditions. Browning also occurs in pure solutions of ascorbic acid, especially at high pH.

The mechanism of ascorbic acid browning is not clearly understood. In citrus juices, browning takes place after the bulk of ascorbic acid has disappeared. The decomposition of ascorbic acid in the presence of air, or under oxidative conditions, atarts with oxidation to dehydroascorbic acid and the transformation of the latter to 2,3-diketo-gulonic acid (see Section 6-4-1). As far as browning is concerned, L-ascorbic acid, dehydro-L-ascorbic acid and 2,3-diketo-1-gulonic acid are interchangeable as starting material in model systems.

The browning mechanism of ascorbic acid–glycine systems in the presence of citric acid was investigated by Lalikainen et al. The rate of pigment formation was considerably greater in the presence of oxygen. Carbon dioxide was produced. Radioactive tracer data indicated that the carbon dioxide evolved was derived mainly from the ascorbic acid and not from glycine. Thus, apparently, Strecker degradation does not occur in this system. The rate of carbon dioxide formation paralleled the rate of browning at 37° C, but not at 50° C. In the presence of oxygen, color intensity reached a maximum, then decreased, as though oxygen had a "bleaching" effect on the pigment.

Furfural (20), *2-furoic acid* (21), *threonic acid* (22), *oxalic acid* (23), an osone of L-xylose (L-erythro-pentosulose, 24), *carbon dioxide* have been characterized as intermediate in the decomposition of ascorbic or dehydro ascorbic acid. Theories regarding the mechanism of their formation and their relationship to the production of dark pigments are mostly speculative.

CHO (20) — COOH (21) — COOH / HCOH / HOCH / CH_2OH (22)

COOH / COOH (23) — O=CH / O=C / HCOH / HOCH / CH_2OH (24)

Ascorbic acid is a reductone and, accordingly, it could enter directly into the state of polymerization or co-polymerization with amines to produce brown pigments. This route, however, does not seem to be responsible for browning in ascorbic acid containing foods. Thus, apparently, decomposition of ascorbic acid must occur first.

10-5. CARAMELIZATION OF SUGARS

As pointed out in Section 10-3, the role of amines and amino acids in the initial stages of the Maillard reaction is that of catalysts. In fact, sugars can undergo browning in the absence of amino compounds, but much higher temperatures are required. The pyrolysis of sugars or *caramelization* is a well known culinary operation for the production of brown color and characteristic flavors.

Although pure sugars caramelize quite rapidly at temperatures above 100° C, several non-amino compounds are known to act as catalysts. These are phosphates, alkalis, acids and salts of carboxylic acids such as citrate, fumarate, tartrate and malate. The exact mechanism of caramelization is not known. It is assumed that the mechanism is similar to that of sugar-amino browning and involves *enolization, dehydration* and *fission.* Wolfrom offers a scheme leading to HMF.

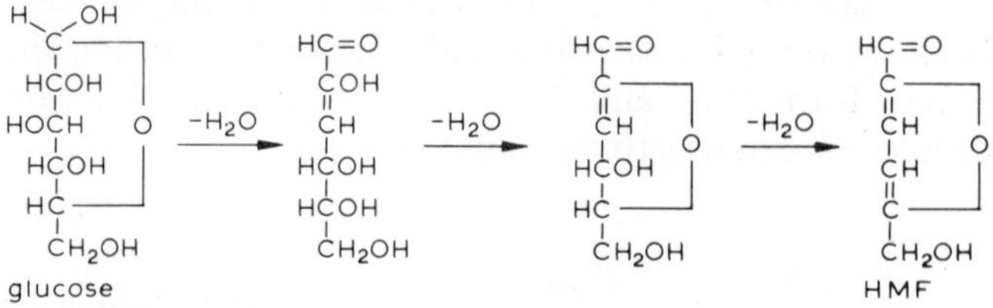

glucose						HMF
H(OH)C		HC=O		HC=O		HC=O
HCOH		COH		C		C
HOCH	$\xrightarrow{-H_2O}$	CH	$\xrightarrow{-H_2O}$	CH	$\xrightarrow{-H_2O}$	CH
HCOH		HCOH		HCOH		CH
HC		HCOH		HC		C
CH_2OH		CH_2OH		CH_2OH		CH_2OH

Hodge compares the Amadori rearrangement of glycosylamine to the Lobry de Bruin-Alberda van Ekenstein transformation of sugars (Section 6-3-1). Both reactions induce enolization at the 1-2 position and just as the former is essential for the progress of browning in sugar–amino systems the latter could well be a key reaction in caramelization in the absence of amino compounds.

10-6. THE HODGE SCHEME

More than twenty years ago, Hodge proposed a scheme summarizing all the reaction mechanisms leading to the browning of foods. Hodge's scheme, which is still the best available summary, is reproduced in Fig. 24.

In this scheme the encircled letters denote the following reactions:

(A) The Maillard reaction.

(B) The Amadori rearrangement.

(C) Dehydration conversions of sugar addition products or of ascorbic acid into reductones or dehydroreductones and into furfural and hydroxymethyl-furfural (HMF), as well as the loss of the amino groups.

(D) Fission products such as acetal, diacetyl, pyruvaldehyde, etc.

(E) Strecker degradation with the loss of CO_2 from the amino acid moiety.

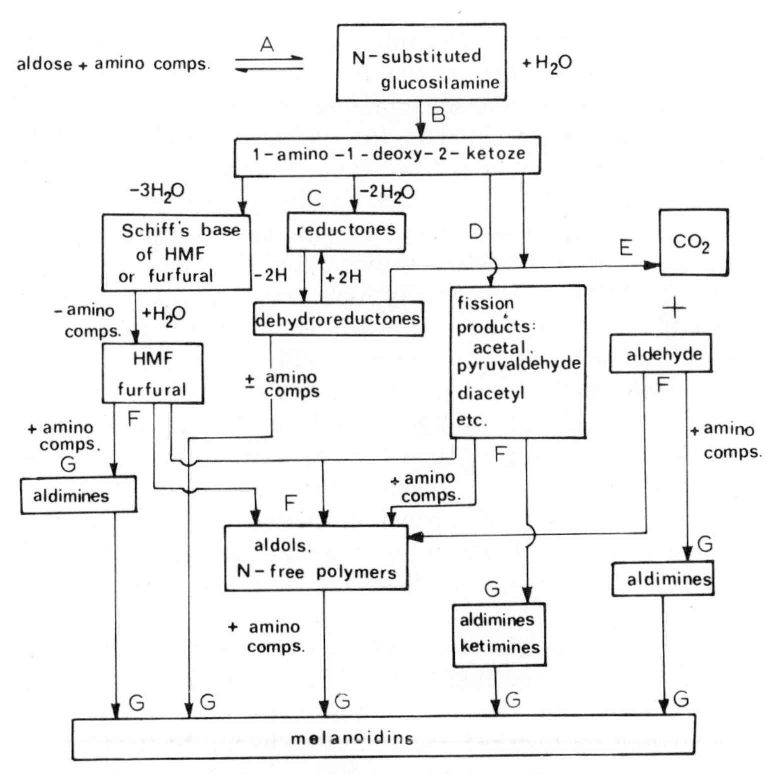

Fig. 24. The Hodge scheme of nonenzymic browning.

(F) Reactions concerning the recombination of amino acid groups or other nitrogenous residues with carbonyl groups and formation of aldimines, etc., or the formation of N-free polymers.

(G) Final polymerization with the creation of melanoidins.

Hodge also includes in his general scheme the catechols and similar phenolic substances which are substrates for enzymic browning, his point of view being that catechols are not different from any other dienolic compound such as ascorbic acid or other reductones.

10-7. EFFECT OF ENVIRONMENTAL FACTORS

The rate of non-enzymic browning is strongly influenced by conditions such as temperature, pH, moisture content, presence of accelerators or inhibitors.

(a) *Effect of temperature.* The rate of browning increases with temperature regardless of the mechanism involved. The activation energies of HMF accumulation, sugar–amine interaction and pigment formation have been determined in various model systems and in actual foods. The values obtained are quite similar and usually fall within the range of 24-26 kcal/mol. Accordingly, the Q_{10} value of non-enzymic browning, in the range of ambient storage temperatures, is about 3. Indeed, severe browning is the main cause of spoilage in foods stored at abnormally high temperatures. It is often observed that foods that had been exposed to higher temperatures in the course of processing darken more rapidly during subsequent storage. A valuable method for predicting the rate of browning of foods is that of *accelerated storage tests.* The food is stored at a temperature 20 or 30° C higher than the expected storage temperature. HMF accumulation or pigment formation is measured periodically and the rate of browning under the experimental conditions is determined. Using a suitable Q_{10} value, the rate of browning under normal storage conditions is computed.

(b) *Effect of pH.* The effect of pH on the rate of browning is more complex. It will be recalled that only unprotonated amine can combine with sugars and therefore the step of sugar–amine condensation is favored by high pH. On the other hand, the subsequent step of rearrangement is acid catalyzed. Not only does the pH affect the rate of browning, but it also determines the predominant mechanism of the process, as we have seen in the case of ascorbic acid browning. In most model systems, faster browning is observed both at very low and very high pH regions with a minimum at some intermediate pH value.

Investigation of the effect of pH on the rate of browning is complicated by the fact that the pH of the system often changes as the browning reactions take place. The mechanism leading to brown pigments in buffered and nonbuffered systems may be different. Care must be exercised in the selection of buffers as alkaline salts of carboxylic acids and phosphates may act as browning catalysts.

(c) *Effect of moisture content.* Within a wide range of water content, the rate of browning increases with increasing solute concentration. Thus, fruit juice concentrates darken much faster than do single strength juices. However, below a certain moisture level, the rate of browning decreases with decreasing water content. Dehydrated foods, with water contents much below their BET monolayer value, do not undergo appreciable browning. Thus, the curve depicting the dependence of browning rate on moisture passes through a maximum.

All chemical reactions occur faster as the concentration of the reactants is increased. However, the effect of moisture content on browning cannot be explained quantitatively by this "dilution effect" alone. Furthermore, this effect does not explain the maximum mentioned above. The mechanism of browning depression at very low moisture is connected with the immobilization of the reactant. Eichner and Karel have shown that if a suitable non-aqueous solvent is present in sufficient quantity, the maximum is suppressed and the increase in browning rate with decreasing water content becomes continuous. This finding also refutes the validity of the "dilution effect", at least in very concentrated systems. Karel explains the browning depressing effect of water by the fact that water is in itself a product of the browning reaction (dehydration of sugar intermediates), and hence an inhibitor.

(d) *Effect of oxygen.* In at least one type of browning, that involving ascorbic acid, oxygen is directly involved in the reaction and its effect on browning is predictable. However, oxygen is not essential for the Maillard reaction. In actual food systems, the presence of oxygen has been found to accelerate browning in some cases, to inhibit browning in others. In some models, the effect of oxygen depends on the moisture content and temperature. This is usually interpreted as a proof of the assumption that temperature and moisture not only affect the rate of browning, but may actually change the dominant mechanism of the reaction. At any rate, in contrast to enzymic browning, exclusion of oxygen or use of antioxidants are not effective in the practical prevention of non-enzymic browning reactions not involving ascorbic acid.

(e) *Browning accelerators and inhibitors. Phosphates, carboxylic acids* and their salts accelerate non-enzymic browning and increase the final color intensity. The effect of metals is not uniform. Copper may be expected to enhance ascorbic acid browning in the light of its effect on the oxidation of this substance. As far as the Maillard reaction is concerned, contradicting observations have been reported on the effect of copper and iron. Tin seems to retard non-enzymic browning in model systems and in foods. For instance, citrus products packed in lacquered cans or glass brown much more readily than those packed in plain tinplate containers. The inhibiting effect of tin, at the low pH of these products, may be due to the strongly reducing conditions prevailing in such containers (hydrogen, stanneous ions). It should be kept in mind that the dominant mechanism of browning in citrus products involves ascorbic acid oxidation.

The best known inhibitor of browning reactions is *sulfur dioxide.* Sulfurous acid and its salts have been long used as browning inhibitors in dried fruits, vegetables and concentrated fruit juices. Several theories have been advanced to explain the mechanism of sulfite action.

(1) *Blocking of carbonyl groups.* Sulfurous acid blocks carbonyl groups through a well known addition reaction

$$R_1R_2C{=}O + H_2SO_3 \longrightarrow HO{-}C(R_1)(R_2){-}OSO_2H$$

It has been postulated that sulfur dioxide prevents aldose–amine interaction by blocking the carbonyl groups of the reducing sugar. Braverman and his co-workers prepared and crystallized such addition products from various sugars. However, these products are quickly hydrolyzed even at moderate water activity levels. Kinetic studies on model systems containing sugars, amino acids, sulfites, do not support the aldose-blocking theory. It is more probable that sulfur dioxide combines with more characteristic carbonylic substances (osuloses, HMF) formed at the later steps of the reaction. Furthermore, sulfites may also interfere with the Strecker degradation by blocking the dicarbonyl compounds entering this reaction.

(2) *Antioxidant effect.* In systems involving ascorbic acid oxidation, sulfur dioxide may be expected to act as an antioxidant, and thus retard pigment formation.

(3) *Bleaching effect.* The earliest theories on the effect of sulfur dioxide on browning speculated that sulfur dioxide might bleach the brown pigments and thus reduce the color intensity. Experimental

evidence does not support this view. Incorporation of sulfur dioxide into systems which are already at an advanced stage of the Maillard reaction (say, after the appearance of HMF) does not prevent browning.

(f) *Practical prevention of browning.* In the strength of the above discussions, it is evident that in such complex systems as foods, it is very difficult to suggest general methods for the prevention of non-enzymatic browning. It will be attempted here only to indicate some possibilities; however, the application of each must be judged separately on its own merits.

(1) Some general precautions can be taken in preventing, or at least greatly decreasing, the rate of browning. The most helpful method is to *refrigerate* the foods subject to this type of change. Lowering the temperature generally slows down all chemical reactions, especially this type of browning.

(2) Similarly, the presence of *sulfites*, as already explained, will hinder the Maillard reaction.

(3) *Lowering the pH* of the product may again be of use if the main cause of browning is the melanoidin condensation. (Dried egg powder, for instance, was once made after the addition of HCl with subsequent admixing of the powder with some Na_2HCO_3 : the excess acid will then combine with the bicarbonate of soda to form NaCl.)

(4) *Lowering the concentration* of the final product sometimes decreases the rate of browning. (Grapefruit and lemon juices are often concentrated at a ratio of only 4 ; 1 instead of 6 : 1, which is usual for orange juice.)

(5) Because the Maillard reaction requires a free carbonyl group in the sugars, it is sometimes possible to prevent browning of certain food products by using *sucrose* instead of reducing sugars, provided, of course, the sucrose will not undergo inversion in the final product during storage.

(6) *Fructose*, as previously mentioned, does not enter into combination with amino acids easily and may probably also be used with the same effect.

(7) When the sugar constitutes only a negligible part of the product, such as is the case with eggs or meat, it can be removed by *fermentation*. (Eggs are sometimes subjected to fermentation before drying them into powder. Similarly, it has been suggested that meat

may be fermented before dehydration; ground meat ($\frac{1}{8}''$) with 5% yeast added is claimed to be well preserved after dehydration.)

(8) Another successful method of removing small quantities of sugar is the use of a mixture of two enzymes, *glucose oxidase* and *catalase* (commercially available under the name of DeeO). The oxidase converts glucose into gluconic acid (which does not combine with the amino groups) and hydrogen peroxide, while the latter is converted into water and oxygen by catalase:

$$R{\cdot}CHO + O_2 + H_2O \xrightarrow[\text{oxidase}]{\text{glucose}} R{\cdot}COOH + H_2O_2$$

$$H_2O_2 \xrightarrow{\text{catalase}} H_2O + \tfrac{1}{2}O_2$$

10-8. FLAVOR FORMATION

One of the most important side effects of non-enzymic browning is the formation of volatile substances. These substances impart to foods characteristic flavors which may be desirable or objectionable. The subject has been reviewed by Hodge.

One of the most significant steps of the browning reaction, as far as flavor generation is concerned, is the Strecker degradation. As we have seen, in this reaction, an amino acid is degraded to an aldehyde of one less carbon atom. Most of these aldehydes are volatile and odorous. Aromas resembling actual food odors are formed in browning model systems. Thus, the flavor of freshly baked bread may be simulated in a system of glucose–leucine when the mixture is heated. Mixtures of sulfur containing amino acids and sugar generate, when heated, meat-like flavors.

Hodge lists some forty volatile substances isolated and identified as products of sugar caramelization. Among those responsible for the characteristic odor of burnt sugars are furfural and its derivatives, *maltol* (25) and *1-methyl-cyclopentenol-2-one-3* (26).

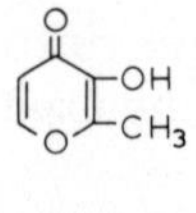

O
OH
CH_3
O

maltol

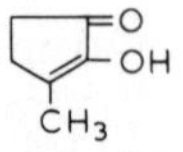

O
OH
CH_3

1-methyl-cyclopentenol-2-one-3

Depending on the relative concentration of the various end products, different flavors may be induced ranging from sweet to pungent or acrid.

10-9. NUTRITIONAL ASPECTS

As mentioned before, one of the results of browning in foods is the irreversible binding of amino acids into complex pigments. These complexes are not hydrolyzed in the digestive tract. Not only the particular amino acid engaged in this type of bond, but also all the amino acids forming the "limit peptide" (see above) are nutritionally lost. Furthermore, amino acids may also be literally *destroyed* through the Stecker degradation.

Of all the amino acids present in proteins, *lysine* is the most susceptible to loss through browning, because of its tendency to combine with carbonyls through its free ϵ-amino group. Since lysine is the limiting essential amino acid in many foods, the effect of such loss on the biological value of the protein is often drastic. The proportion of lysine residues with intact ϵ-amino groups (*available lysine*) may be determined by reacting the food with *dinitro-fluorobenzene.* All the free amino groups combine with the reagent. Following the reaction the protein is hydrolyzed and the amount of *dinitro-phenyl-lysine* in the hydrolysate is determined. Good correlation is observed between the "available lysine" content and nutritional value determined by feeding tests. "Available lysine" is therefore a useful index for the evaluation of heat or storage damage to proteins.

SELECTED BIBLIOGRAPHY TO CHAPTER 10

Clark, A.V. and Tannenbaum, S.R., Isolation and Characterization of Pigments from Protein–Carbonyl Browning Systems, J. Agr. Food Chem., **18** (1970), 891.

Ellis, G.P., The Maillard Reaction, Advances in Carbohydrate Chemistry, **14** (1959), 63-134.

Harris, R.S. and Loesecke, H. Von (Eds.), Nutritional Evaluation of Food Processing, John Wiley & Sons Inc., New York, 1960.

Hodge, J.E., Dehydrated Foods. Chemistry of Browning Reactions in Model Systems, J. Agr. Food Chem., **1** (1953), 928.

Hodge, J.E., Origin of Flavor in Foods, Nonenzymatic Browning Reactions, in Chemistry and Physiology of Flavors, H.W. Schultz, E.A. Day and L.M. Libbey (Eds.), The Avi Publishing Co. Inc., Westport, Conn., 1967.

Karel, M., Recent Research and Development in the Field of Low-Moisture and Intermediate-Moisture Foods, C.R.C. Critical Reviews in Food Technology, **3** (1973), No. 3, 329-373.

Reynold, Thelma M., Nonenzymic Browning Sugar–Amine Interactions, in Carbohydrates and their Roles, H.W. Schultz, R.F. Cain and R.W. Wrolstad (Eds.), The Avi Publishing Co. Inc., Westport, Conn., 1969.

Stadtman, E.R., Nonenzymatic Browning in Fruit Products, Adv. Food Research, **1** (1948), 325-372.

CHAPTER 11

LIPIDS

11-1. DEFINITIONS

Of the three most important groups of nutrients: carbohydrates, proteins and fats, the latter belong to a large class of substances called lipids. The definition of the term "*lipid*" is a subject of controversy. Older texts define lipids as natural substances which are insoluble in water, but soluble in non-polar solvents such as hexane, benzene, carbon tetrachloride, ether, etc. Such a definition, based on solubility alone, encompasses a very large number of very diverse substances. Nevertheless, this definition has been retained by many biochemists, mainly because these substances, *chemically* diverse as they may be, have many common points as far as their biological activities are considered.

Another definition, which we shall adopt in this text, limits the applicability of the term to "*actual or potential derivatives of fatty acids or closely related substances*" (Chapman). So defined, the group includes fats, waxes and related substances. Other water insoluble groups such as terpenes and carotenoids will be discussed in the subsequent chapters.

11-2. FATTY ACIDS

The essential components of lipids are aliphatic carboxylic acids, known as *fatty acids*. These are divided into two main groups, as *saturated* and *unsaturated*. With extremely rare exceptions, natural fatty acids contain an even number of carbon atoms, ranging from 4 to 28 in the ones most commonly found in fats. Fatty acids with a higher number of carbon atoms are found in waxes. The chain is usually straight, unbranched and unsubstituted, although OH substitution, cyclic configurations and branching are not uncommon.

11-2-1. *Saturated fatty acids*. A list of some of the saturated fatty acids is given in Table 3. The common names of these acids indicate the specific source in which they are especially abundant or from

TABLE 3

Common normal saturated (alkanoic) fatty acids

Number of carbon atoms	Common name	M. pt. °C	Formula
4	Butyric acid	−7.9	$CH_3(CH_2)_2COOH$
6	Caproic acid	−3.4	$CH_3(CH_2)_4COOH$
8	Caprylic acid	16.7	$CH_3(CH_2)_6COOH$
10	Capric acid	31.6	$CH_3(CH_2)_8COOH$
12	Lauric acid	44.2	$CH_3(CH_2)_{10}COOH$
14	Myristic acid	53.9	$CH_3(CH_2)_{12}COOH$
16	Palmitic acid	63.1	$CH_3(CH_2)_{14}COOH$
18	Stearic acid	69.6	$CH_3(CH_2)_{16}COOH$
20	Arachidic acid	75.3	$CH_3(CH_2)_{18}COOH$
22	Behenic acid	79.9	$CH_3(CH_2)_{20}COOH$
24	Lignoceric acid	84.2	$CH_3(CH_2)_{22}COOH$
26	Cerotic acid	87.7	$CH_3(CH_2)_{24}COOH$

which they have been isolated. Despite attempts to adopt a more descriptive nomenclature, in accordance with the Geneva System, trivial names are used extensively.

Physical properties vary with the number of carbon atoms, as in any homologous series. Acids with fewer than 12 carbon atoms are conventionally called "*the volatile fatty acids*" since they can be steam distilled with relative ease. Members with carbon atom numbers higher than 10 are solid at room temperature. Solubility in water decreases with chain length and acids with more than 10 carbons are practically water-insoluble.

11-2-2. *Unsaturated fatty acids.* The majority of oils from plant sources contain unsaturated fatty acids. This group also consists, generally, of straight-chain fatty acids with an even number of carbon atoms from C_{10} to C_{24}. The possibilities for isomers existing among them are largely due to: (a) the number of unsaturated double bonds (mono-,di-, tri- and tetraethenoid) present; (b) their position in the chain; and, (c) the possibility of cis or trans configurations.

Here are a few of the most important unsaturated fatty acids and their sources:

Oleic acid (octadeca-9-enoic)

$CH_3(CH_2)_7CH{=}CH \cdot (CH_2)_7 \cdot COOH$, in olive oil.

Linoleic acid (octadeca-9-12-dienoic)

$CH_3(CH_2)_4CH{=}CH \cdot CH_2 \cdot CH{=}CH(CH_2)_7 \cdot COOH$ in linseed oil and soybean oil.

Linolenic acid (octadeca-9-12-15-trienoic)

Arachidonic acid (eicosa-5-8-11-14-tetraenoic)

Erucic acid $CH_3(CH_2)_7CH{=}CH(CH_2)_{11}COOH$ rapeseed.

According to world statistical figures, two of the unsaturated fatty acids, oleic and linoleic, account for 34% and 29%, respectively, of all the edible oils produced by man annually, as against only 11% for palmitic acid, a saturated fatty acid. Unsaturated fatty acids carrying triple bonds are uncommon.

Acids with two or more double-bonds are known as "*poly-unsaturated acids*" (e.g., linoleic, linolenic and arachidonic acids). Polyunsaturated fatty acids perform certain important physiological functions, but they cannot be synthesized in the body fast enough and must be supplied in the food. They are sometimes referred to as "essential fatty acids". Linoleic acid is the most abundant member of this group.

The unsaturated fatty acids have considerably lower melting points than the corresponding saturated fatty acid. Thus oleic acid, with 18 carbon atoms, is a liquid at room temperature. The industrial hardening of fats by hydrogenation is based on the saturation of the double bonds in unsaturated fatty acid residues.

11-2-3. *Cyclic, branched and substituted fatty acids.* Cyclic configurations are found in a few naturally occurring fatty acids. *Malvalic acid,* a constituent of cottonseed oil and oils from other malvaceous plants, contains a cyclopropenyl group close to the center of the molecule. This acid causes several disturbances in hens and must be removed from cottonseed oil intended for use in poultry feeding.

$$CH_3(CH_2)_6C{=}C-(CH_2)_7-COOH \quad (\text{ring: } C{=}C \text{ bridged by } CH_2)$$

malvalic acid

Cyclopentyl groups are found in the so-called *chaulmoogra acids.*

Branched fatty acids have been found in the body oils of dolphin and porpoise as well as in mutton fat, wool fat, butter fat and in some lipids of bacterial origin. Usually the chain carries a single side group, consisting of methyl. The occurrence of isovaleric (β-methyl butyric) acid in natural fats has been established long ago. Of the hydroxyl substituted fatty acids, *ricinoleic acid* is the best known. It constitutes 80-85% of castor oil. Because of the reactivity of the hydroxyl group, ricinoleic acid was in high demand as a raw material for the synthesis of certain plastic polymers.

$$CH_3(CH_2)_5-CH-CH_2-\underset{\displaystyle OH}{\underset{|}{C}}H=CH-(CH_2)_7COOH$$

ricinoleic acid

11-2-4. *Isomerism in fatty acid.* From the structure of both saturated and unsaturated fatty acids, it is evident that there can be three possible types of isomerism in these compounds:

(a) The single isomerism of a straight chain versus a branched chain, as, for instance, in butyric and isobutyric acids:

$CH_3 \cdot CH_2 \cdot CH_2 \cdot COOH$ — butyric acid

$(CH_3)_2 > CH \cdot COOH$ — isobutyric acid

(b) Isomerism caused by the position of the double bond in the chain of an unsaturated fatty acid, as for instance in oleic acid and isooleic acids:

oleic acid - $CH_3 \cdot (CH_2)_7 CH=CH(CH_2)_7 \cdot COOH$

isooleic acid - $CH_3(CH_2)_4 CH=CH(CH_2)_{10} \cdot COOH$

In the case of more than one unsaturated double bond, this type of isomerization can give two distinct kinds of systems, conjugated and nonconjugated:

$-C=C-C=C-$ — conjugated

$-C=C-C-C=C-$ — non-conjugated

(c) The third type of isomerism which can occur in unsaturated fatty acids is cis-trans (geometrical) isomerism, as in the case of C_{18} mono-unsaturated fatty acids:

$$\begin{matrix} CH_3(CH_2)_7 \cdot CH \\ \| \\ HOOC \cdot (CH_2)_7 \cdot CH \end{matrix} \qquad \begin{matrix} CH_3(CH_2)_7 \cdot CH \\ \| \\ CH \cdot (CH_2)_7 \cdot COOH \end{matrix}$$

oleic acid *(cis)* — elaidic acid *(trans)*

In nature, most unsaturated fatty acids occur mainly in the cis-form, while only traces of trans-forms have been detected in natural lipids. However, these forms have been found in relatively large amounts in the body fats of ruminants as well as in natural lipids which have been subjected to hydrogenation.

The steric configuration of fatty acids has a marked effect on the physico-chemical properties of fats. The cis-configuration introduces a bend of approximately 30° in the chain, which is somewhat shortened by this effect. The trans-configuration results in spatial arrangements very close to those of saturated fatty acids. This is reflected by differences in melting point. The melting point of oleic acid is 30° C lower than that of the transisomer, *elaidic acid.*

11-3. FATS AND OILS

11-3-1. *Composition.* The all important group of foods, known as fats and oils, are *glycerides*, i.e. glycerol esters of fatty acids. In fats and oils, all the three hydroxyls of the glycerol molecule participate in ester bonds, hence their chemical name of *triglycerides*, or *neutral glycerides*:

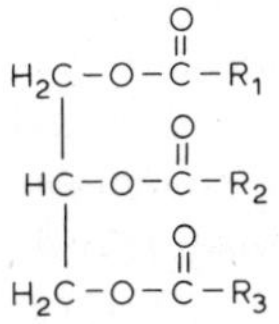

$$
\begin{array}{l}
H_2C-O-\overset{O}{\overset{\|}{C}}-R_1 \\
\;| \\
HC-O-\overset{O}{\overset{\|}{C}}-R_2 \\
\;| \\
H_2C-O-\overset{O}{\overset{\|}{C}}-R_3
\end{array}
$$

triglyceride

It is merely an accepted convention to designate by the name "*fats*" only these glycerides which will solidify at ordinary room temperature, while the name "*oils*" is assigned to all others, which will remain liquid in these circumstances. The structures of the component fatty acids are really the determining factor: the more saturated the fatty acids, the higher the melting point of the fat. All natural fats and oils are mixtures of triglycerides in which the three fatty acids esterifying the glycerol usually differ from each other: nature's principle in this case is maximum heterogeneity of the fatty acids (F) on one glycerol (G):

$$GF_1F_2F_3$$

Less common is the combination in which two fatty acids are identical $GF_1F_1F_2$. The case in which all three hydroxyls are esterified by the same fatty acid is very rare indeed.

The pattern of fatty acid distribution among the two different positions of esterification on the glycerol molecule (i.e. the primary and secondary alcohol positions) is not entirely at random, nor entirely even. In vegetable fats the 2-position is the preferred site of esterification with unsaturated residues while the saturated acids occupy mainly the 1 and 3 positions. Preferential distribution patterns are less clear in animal fats.

A large number of fat soluble substances accompany the triglycerides in natural oils and fats. These include pigments (chlorophylls, carotenoids), oxidation products (aldehydes, ketones), free fatty acids, sterols, etc. In edible oils, the major proportion of these components is removed in the process of refining (see further).

Until recently, the study of the composition of natural triglycerides was a difficult task. The subject involves two questions: the fatty acid composition and the distribution of the fatty acid among different triglyceride species. In order to determine the fatty acid composition, the fat is first hydrolyzed (saponified, see Section 11-3-3) by alkali. The fatty acids are liberated from their salts (soaps) by acidification. The determination of the different fatty acids in the mixture by means of chemical reactions is difficult, because of the chemical similarity of most acids. Accurate specific reactions exist for only a few acids. Physico-chemical techniques are more adequate. The most convenient methods are *gas-liquid chromatography* (GLC) and *thin-layer chromatography* (TLC). For the purpose of GLC analysis, the fatty acids in mixture must be first converted into more volatile derivatives, usually methyl esters.

An approximate idea of the fatty acid composition of fats may be obtained by a number of practical tests. The amount of alkali consumed (in mg of KOH), for the total hydrolysis of one gram of fat (*saponification number*) is an indication of the average molecular weight of the fatty acids. A fat containing mostly smaller fatty acids will show a higher saponification number. The amount of halogen (as g. of iodine) which can be fixed by 100 grams of oil (*iodine number*) is proportional to the abundance of unsaturated bonds. Short fatty acids can be determined after separation by steam distillation or extraction with water. All these tests are used for gross characterization of edible fats and oils.

11-3-2. *Physical properties.* Pure triglycerides are colorless, tasteless, odorless, water insoluble substances. Any color, odor and taste in fats and oils are due to non-triglyceride components mentioned before.

The solid-liquid transition of fats is of considerable technological importance. Triglycerides containing a large proportion of unsaturated fatty acids have lower melting points. Most vegetable oils are therefore liquid at room temperature. Glycerides can exist in a number of different crystalline forms (*polymorphism*), hence the phenomena of multiple phase transition (melting) points. The crystalline characteristics and melting behavior of fats do not depend only on fatty acid composition but also on the distribution of fatty acids among the triglyceride molecules. Thus, a mixture of different oils and fats may be prepared, so as to have the same *fatty acid composition* as cocoa butter, but the consistency and melting behavior of such a mixture would be very different from that of cocoa butter. However, if such a mixture is subjected to a process of *transesterification*, i.e. controlled fatty acid interchange between glyceride molecules, a product with new physical properties is obtained. Transesterification is the basic process in the manufacture of "*tailor-made*" fats such as imitation cocoa butter substitute.

In most food systems, fats and oils occur as dispersed phases (droplets or globules), stabilized by a layer of emulsifying substances such as proteins, phospholipids, mono- and diglycerides.

11-3-3. *Hydrolysis.* Glycerides are easily cleaved into fatty acids and glycerol by heating in alkali. The resulting alkaline salts of fatty acids are the well known "*soaps*", hence the name of "*saponification*" given to hydrolytic cleavage of fats, and by extension to the hydrolysis of all esters.

The de-esterification of triglycerides is also catalyzed by the enzyme *lipase.* This enzyme is widespread in all lipid-containing tissues. Lipases from different sources show varying degrees of specificity. In the case of lipases, specificity may involve selectivity towards different fatty acids as well as preference towards the position of the ester bonds on the glycerol backbone. Thus, the enzyme responsible for the digestion of fats, pancreatic lipase, has some preference for the 1, 3 positions and for shorter chain fatty acids. As ester bonds are gradually broken, intermediate products, di- and monoglycerides are formed.

F_1		OH	F_1		OH	OH	F_1		OH
F_2	$\xrightarrow{HOH}$	F_2	F_2	$\xrightarrow{HOH}$	OH	F_2	OH	$\xrightarrow{HOH}$	OH
F_3		F_3,	OH		F_3,	OH,	OH		OH
triglyceride		diglycerides			monoglycerides				glycerol

Lipases react in heterogenous systems such as emulsions of glycerides in aqueous media. Their action occurs at the interface between the phases. From the technological point of view, the most significant consequence of lipase activity in foods is the development of a harsh, acrid taste as a result of free fatty acid liberation. The short chain volatile fatty acids (such as butyric acid) also contribute their characteristic odor to foods so affected. This type of deterioration, known as *hydrolytic rancidity*, is quite common in olives, milk, cream, butter and nuts. As lipase activity occurs on the interphase, hydrolytic rancidity is more rapid in more finely dispersed emulsions, such as homogenized milk and cream. Lipase catalyzed hydrolysis of oil is a serious problem in the manufacture of olive oil. As a result of this reaction the free fatty acid content of olives increases rapidly during storage, especially if the olives have been bruised. A high free fatty acid level in edible oils is objectionable and these must be removed in the process of refining, which represents substantial losses in yield. Hydrolytic rancidity is not always objectionable. The characteristic flavor of "blue" cheeses is due mainly to short chain fatty acids formed by the action of mold lipases on butter fat.

The intermediate products of fat hydrolysis, the mono- and diglycerides are valuable in food industries as emulsifying agents. They are not produced by partial saponification of fats but rather by controlled partial esterification of glycerol with fatty acids. They are extensively used in the stabilization of emulsions, such as margarine, salad dressings, butters, coffee creamers, etc.

11-3-4. *Oxidation.* The tendency of fats and oils to become rancid is a phenomenon well known since ancient times. Hydrolytic rancidity has been described above. A much more frequent and serious type of rancidity results from oxidative reactions. The susceptibility of fats and fatty acids to oxidation is associated with the presence of unsaturated bonds. The spontaneous nonenzymic oxidation of lipids exposed to air is termed *autoxidation.* This is the most frequent type of oxidative deterioration of lipids in manufactured food products. On the other hand, in vegetables, lipid oxidation may be catalyzed by the enzyme lipoxidase. In meat and fish, lipid oxidation is catalyzed by hematin compounds. In view of its importance in food industries, oxidative deterioration of lipids will be treated in more detail in Chapter 15.

11-3-5. *Manufacture and processing of edible oils.* Oils from plant materials are obtained by two general methods: *pressing* and *solvent extraction.* Oilseeds which have a plastic consistency and contain a relatively high percentage of oil are preferentially processed by the first method, which is more economical. The seeds are cleaned, crushed and decorticated if necessary, then cooked and pressed. The purpose of cooking is to denature the proteins which would otherwise stabilize the oil–water emulsion, to decrease the viscosity of the oil, to cause coalescence of fat globules and to increase the plasticity of the seeds. The moisture level is carefully controlled. Usually, moisture is added before cooking, then removed at the final stages of the operation. The cooked seeds are now subjected to very high pressure, in the order of a few hundred atmospheres, in continuous screw presses or expellers. As a result of friction in the expeller, very high temperatures are reached and considerable thermal damage occurs both in the oil and in the residual cake. The recovery of oil by pressing is not complete and the pressed cake contains 4 to 8% oil, which may be recovered by solvent extraction. In the so-called pre-press solvent extraction processes, the seeds are pressed mildly to a cake with 10-16% residual oil, which is then solvent extracted. Excessive heat damage in the expeller is thus avoided.

Soybean oil is manufactured almost exclusively by solvent extraction. The seeds are tempered (heated to increase plasticity), and flaked between rolls to increase the surface area. The flakes are extracted with a suitable non-polar solvent, in counter current fashion. The most commonly used solvent is a light petroleum fraction, commercially known as "hexane". The solution of oil in solvent (miscella) is distilled to recover the solvent and the oil is sent to the refinery. The spent flakes are also desolventized and dried.

The next steps in the manufacture of edible oils are concerned with refining the crude oil. The purpose of these steps is to remove a large proportion of the non-triglyceride components. Free fatty acids are removed by neutralization with alkali. Phospholipids (gums) are removed by addition of dilute phosphoric acid. The gums absorb water, swell and form a separable phase. The pigments are eliminated by special absorbents such as Fuller's earth. The final step is that of deodorization whereby relatively volatile components (aldehydes, ketones, etc.), are removed by vacuum-steam distillation at temperatures of 200-250° C. The refined oil is immediately cooled and filtered.

Fats are recovered from animal tissue by rendering (tallow), pressing or solvent extraction (fish liver oil).

As mentioned before, natural vegetable oils are usually liquid at ordinary room temperature because they contain a large proportion of unsaturated fatty acids in their triglycerides. However, it was shown, long ago, that if these could be converted into saturated fatty acids the oils would become solid fats. Sabatier was the first to show that vegetable oils could be hydrogenated by gaseous hydrogen in the presence of a catalyst such as metallic nickel. This process is the essential unit operation in the manufacture of hardened oils, used in the production of margarine and shortenings.

Commercial hydrogenation is usually carried out in large vessels under a pressure of 2-10 atmospheres and at temperatures between 110° and 119° C with finely divided suspensions of nickel at concentrations of 0.03-0.10%, based on the weight of oil. The reaction is halted when the product has reached an iodine number in accordance with the desired consistency of the fat.

During this reaction hydrogen is added to the double bonds of the hydrocarbon chains of the unsaturated fatty acid groups in the triglyceride molecules, so that linolenic acid may be converted into linoleic or isolinoleic acid, linoleic to oleic acid and oleic into stearic acid. However, it has been shown that the double bonds farthest removed from the ester linkage are more reactive than those nearest to it. Moreover, the hydrogenation reaction tends to be selective in the sense that the greater the degree of unsaturation of a fatty acid group the greater is its tendency to add hydrogen. As a consequence of this selectivity, a fatty acid, such as linoleic, for instance, having two double bonds, will tend to be hydrogenated before an oleic acid group, attached to the same glyceride, because it has only one double bond. The course of hydrogenation can be controlled, however, by varying the process conditions (type of catalyst, pressure, temperature). Controlled hydrogenation processes permit hardening of oils to the desired extent while at the same time minimizing the loss of polyunsaturated fatty acids.

A change in melting characteristics is not the only effect of hydrogenation. Since double bonds are the most reactive function in fats, especially towards oxidation, their saturation with hydrogen increases the chemical stability of the product.

Extensive isomerization occurs as a result of hydrogenation. Several types of isomerization are possible:

(a) Cis → trans transformation of unsaturated fatty acids, simply as a consequence of exposure to high temperature.

(b) Formation of "unnatural" fatty acids by saturation of the "wrong" double bond in polyunsaturated residues (e.g. the 9 : 10 double bond of linoleic acid to give an "iso"-oleic acid).

(c) Migration of double-bonds.

The unnatural trans-isomers are thought to induce undesirable cytological changes. However, the toxicological significance of trans-isomerization in practical oil hardening is not yet clear.

11-4. WAXES

As mentioned before, waxes are a type of lipid, in which the fatty acids are esterified by higher monohydric alcohols rather than by glycerol. These alcohols contain from 24 to 36 carbon atoms. Natural waxes are mixtures of many such esters and often contain unesterified alcohols, ketones and hydrocarbons having an odd number of carbon atoms. These hydrocarbons are apparently formed from long chain fatty acids by a process of decarboxylation. Waxes serve mainly as protective, water repellant coatings on the surface of tissues and organisms. Their function is to prevent undue evaporation of moisture or invasion of the tissue with water from the environment.

Some commercially important natural waxes are: *beeswax* — secreted by the honeybee and containing, among other constituents, palmitic and cerotic acids and melissyl alcohol; *carnauba wax* — which coats the leaves of the carnauba palm, and which contains even higher alcohols and fatty acids; *lanolin* — a waxy material obtained from wool; *spermaceti* — found in the head of the sperm whale and containing cetyl and oleyl alcohols and palmitic acid.

As a rule, the waxes are solids with melting points between 60° and 80° C; they are more resistant to saponification than fats and oils, and are also less susceptible to auto-oxidation.

11-5. PHOSPHOLIPIDS

This third group which consists of lipids containing phosphoric acid radical is difficult to classify in view of its wide heterogeneity.

However, one sub-group, the phosphoglyceride (also termed glycerophospholipids or glycerol phosphatides) is the most important. The general structure of a phosphoglyceride is as follows:

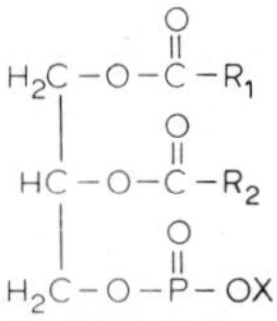

$$
\begin{array}{l}
H_2C-O-\overset{\overset{O}{\parallel}}{C}-R_1 \\
\;\;| \\
HC-O-\overset{\overset{O}{\parallel}}{C}-R_2 \\
\;\;| \\
H_2C-O-\overset{\overset{O}{\parallel}}{P}-OX
\end{array}
$$

Two hydroxyls of the glycerol residue are esterified with fatty acids. The third hydroxyl is bound to phosphoric acid which in turn is ester-linked with X—OH, usually an amine alcohol.

The phosphoric acid end of the molecule is strongly polar, hydrophylic while the fatty acid "tails" are non-polar. This dual structure (sometimes termed *amphiphathic*) makes the phosphoglycerides valuable surface active agents and emulsion stabilizers. Phosphoglycerides are important constituents of the cell wall. Many biological disorders involving cell wall malfunctions are due to irregularities in phospholipid deposition.

Lecithin (choline phosphoglyceride or phosphatidyl-choline) is a widely distributed phospholipid found in animal tissues, egg yolk and seeds. Soybean lecithin is produced commercially and used mainly as an emulsifying agent in chocolate and other foods.

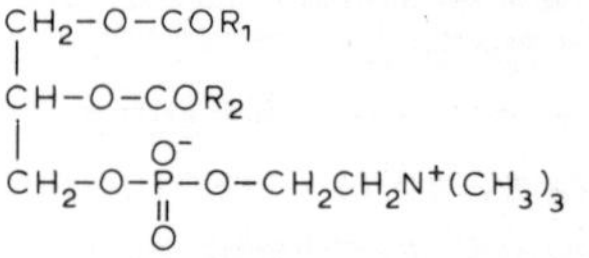

$$
\begin{array}{l}
CH_2-O-COR_1 \\
| \\
CH-O-COR_2 \\
| \\
CH_2-O-\underset{\underset{O}{\parallel}}{\overset{O^-}{\overset{|}{P}}}-O-CH_2CH_2N^+(CH_3)_3
\end{array}
$$

lecithin

Lecithin from various sources contains different saturated or unsaturated fatty acid groups. As shown above, lecithin is amphoteric. At pH 7, it forms a zwitterion, a dipolar ion in which the negative charge on the phosphoric acid residue is neutralized by a positive charge on the quarternary nitrogen of choline. The polar nature of the phosphoric acid–choline residue activates the entire molecule. Lecithins, as well as other phospholipids, are easily oxidized or hydrolyzed and combined with a number of other substances such as proteins and carbohydrates.

From the structural formula of lecithin it is clear that two positional isomers exist, depending on position of the phosphate groups (esterification of the primary of secondary OH of glycerol). Both isomers occur naturally.

Cephalins are phosphatides found in eggyolk and also in many animal tissues, particularly in nervous tissue. The main component of "cephalin" is *phosphatidyl ethanolamine* in which X—OH is ethanolamine. Another wide distributed phosphatide is *phosphatidyl serine* in which X—OH is the amino acid serine:

$$HO{-}CH_2{-}\underset{\displaystyle NH_2}{\underset{|}{C}H}{-}COOH$$

This substance is a minor constituent of "cephalin". Other phosphoglycerides containing glycerol, inositol, inositol phosphates and glucose as their X—OH group are also known.

Phospholipids not derived from glycerol are less abundant but of considerable biological importance (*sphingolipids* and *glycolipids*).

Phosphoglycerides are hydrolyzed by alkali or specific enzymes. Mild alkaline hydrolysis of lecithin saponifies the fatty acids, leaving a *choline glycerophosphate.* Stronger alkali also cleaves the bond with choline. *Phospholipases* vary in their specificity. *Phospholipase* A, found in snake venom, hydrolyzes the bond with the fatty acid at 2 positions. The product of the reaction is *lysolecithin.* Lysophosphatides cause lysis (disintegration) of cell membranes and are highly toxic. *Phospholipase* B removes both fatty acids, and the result is similar to that of mild alkaline hydrolysis. The bond between glycerol and phosphoric acid is hydrolyzed by *phospholipase* C, yielding a diglyceride. *Phospholipase* D cleaves the bond with X—OH, to give a phosphotidic acid.

Phospholipids (as well as other lipids with polar groups, such as fatty acids and soaps) disperse in water in an orderly way whereby the non-polar "tails" are hidden from the aqueous medium. Several types of organized structures are then possible, such as thin, monomolecular *films* on the surface, *micelles* inside the aqueous medium and *bimolecular layers* between two regions of the aqueous medium (Fig. 27). Phosphoglyceride bilayers are about 70 Å thick. These structures are very interesting as they have transport properties similar to those of biological membranes. They are permeable to water but retain simple inorganic ions. This fact, together with the occurrence of phospholipids as cell wall components, led to the current assumption that a structure consisting of a lipid bilayer attached on both polar faces to protein is an acceptable general model of biological membranes.

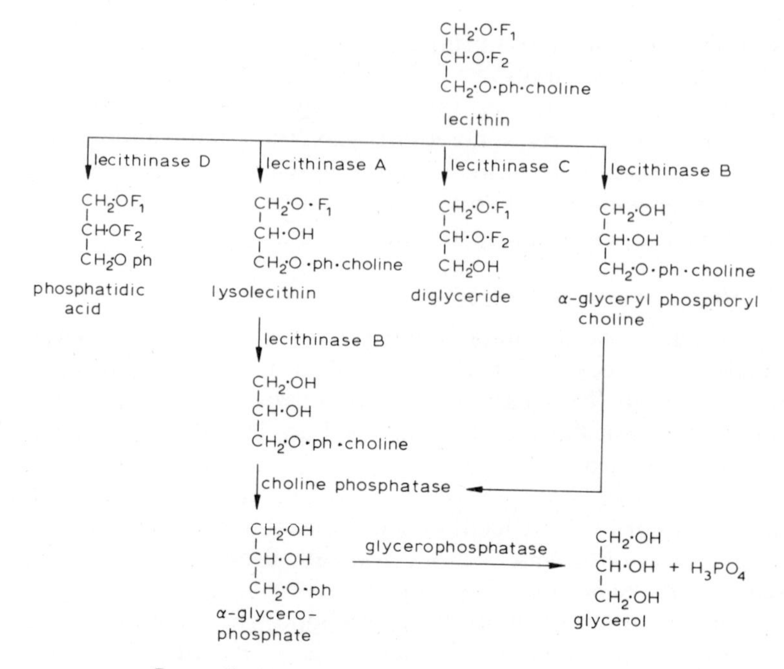

Enzymatic cleavage of lecithin

11-6. ROLE OF LIPIDS IN FOODS

Nutritionally, the main function of fats is to supply energy. The nutritional caloric content of fats is very high, 9 cal/g, as compared with approximately 4 for carbohydrates or proteins. In addition, dietary fats are important as vehicles for fat soluble vitamins and as a source of essential fatty acids. In recent years, the physiological aspects of lipid nutrition has been a field of immense activity, especially in connection with the relationship between dietary fat intake and heart diseases, atherosclerosis, blood cholesterol, etc.

The biological role of fats "in vivo" is complex. Fats are the preferred form of long-term storage fuel in the living organism, but in addition many lipids, by virtue of their special surface properties, are important structural elements of biological membranes, as explained before. Fats and waxes serve as protective surface coatings on plant leaves, insects, and cell walls of a number of microorganisms.

As far as the sensory characteristics of foods are concerned, the role of lipids is mainly connected with texture and rheological properties. Although neutral, refined fats have no taste of their own, the presence and physical form (dispersion) of fats determine, to a large extent, the taste sensation and mouth feel received from food. This is again due to the fact that in many food systems the flow properties of the food in the mouth (spreadability, coating of the tongue, sensation on swallowing, viscosity) are controlled by the fat fraction. Many foods (milk, cheese, soups, pastes, creams, ice cream) taste "richer" when they have a high proportion of finely dispersed fat.

The organoleptic consequences of lipid autoxidation (off-flavors, browning, effect on protein toughness, etc.), will be discussed in Chapter 15.

SELECTED BIBLIOGRAPHY TO CHAPTER 11

H.A. (Ed.), Analysis and Characterization of Oils, Fats and Fat Products, Vols. 1 and 2, Interscience Publishers, New York, 1964 and 1968.

Chapman, D., Introduction to Lipids, McGraw Hill, London, 1969.

Gur, M.I. and James, A.T., Lipid Biochemistry, an Introduction, Cornell University Press, Ithaca, New York, 197, .

Holman, R.T. (Ed.), Progress in the Chemistry of Fats and Other Lipids, Vols. 1 to 12, Pergamon Press, London, 1952–1972.

Markley, K.S. (Ed.), Fatty Acids, their Chemistry, Properties, Production and Uses, 2nd Edn., Vols. 1 to 4, Interscience Publishers Inc., New York, 1960–1967.

Swern, D. (Ed.), Bailey's Industrial Oil and Fat Products, 3rd Edn., Interscience Publishers, New York, 1964.

CHAPTER 12

CAROTENOIDS

The yellow companions

R. WILLSTAETTER

12-1. OCCURRENCE

The *carotenoids* are a large group of pigments, widely distributed in the plant and animal kingdoms. The carotenoids are yellow-orange to purple in color, insoluble in water, but soluble in fats and organic solvents; they are classed, therefore, as *lipochrome pigments.* More than 300 individual carotenoids are known today.

The carotenoids embrace two groups:

(a) *Carotenes* — hydrocarbons, soluble in petroleum ether but only slightly soluble in ethanol.

(b) *Xanthophylls* — oxygenated derivatives of the carotenes; these compounds are alcohols, aldehydes and acids, soluble in ethanol, methanol and petroleum ether.

In the higher plants the carotenoids are found in the leaves, together with chlorophyll, as well as in many other parts of the plant. Their association with chlorophyll has earned them the title of "*the yellow companions*". They are found in carrots (from which the name *carotene* was derived), and in sweet potatoes; these generally accumulate hydrocarbons (β-carotene) rather than xanthophylls. They are also found in fruits, e.g. tomatoes, peaches, all citrus fruits, banana skins, squashes, paprika, red peppers, rose-hips and many others.

Fruits containing carotenoids may be divided into four main categories:

(a) Those containing in their plastids small concentrations of normal "plastid carotenoids", in addition to chlorophyll *a* and *b*, e.g. elderberries.

(b) Those in which ripening causes a marked synthesis of acyclic carotenoids, such as lycopene, but little β-carotene, e.g. tomatoes, watermelons and apricots.

(c) Fruits in which the main pigment is carotene, of a mixture of xanthophylls accompanied by only small amounts of acyclic carotenoids, e.g. red palm, citrus fruit.

(d) Fruit in which the main constituents of the carotenoid pigment are specific xanthophylls such as *capsanthin*, while both cyclic or acyclic carotenes are only minor components.

Carotenoids constitute, also, the principal pigments of certain yellow, orange and red flowers, and of many microorganisms (red and green algae, some yeasts, fungi and photosynthetic bacteria). They are also found in all animals. While plants and microorganisms synthesize their carotenoids, those present in the tissues of higher animals are derived from dietary sources. Food products of animal origin such as milk, butter, egg yolk, some fish and shellfish, contain carotenoids dispersed in the lipid components.

Synthetic carotenoids are often added to foods as colorants. An interesting carotenoid, *astaxanthin*, 3:3′-dihydroxy-4:4′-diketo-β-carotene constitutes the red principle in shrimp, lobster and salmon.

12-2. STRUCTURE

Carotene was first extracted from carrots in 1831 by Wackenroder, but another 100 years elapsed before its structure was definitely established by Karrer in 1930. All carotenoids belong to the class of "polyenes" in organic chemistry, namely, long chains of conjugated double bonds. The presence of many conjugated unsaturated bonds explains the intense color of carotenoids ranging from yellow to red and purple.

The second consideration to be taken into account regarding the structure of all carotenoids is their isoprenic nature. Similar to a number of other groups of biologically important substances, carotenoids are built of isoprene units:

$$CH_2{=}\underset{\displaystyle CH_3}{\underset{|}{C}}{-}CH{=}CH_2$$

isoprene

It is also obvious from the structure of carotenoids that in unsaturated compounds of this type a very large number of geometrical isomers of the cis and trans configurations are possible. However, as has been shown by Zechmeister and his collaborators, to whom we owe most of our present knowledge of carotenoids, the bulk of carotenoids found in nature appear to have an all-trans configuration and only very few cis isomers are known.

The carotenoids are characterized by the possession of a bilaterally symmetrical C_{40} skeleton, each half of which can be considered from the formal structural stand-point to consist of four isoprene units joined head to tail, so that the projecting methyl groups assume a 1:5 relationship. These two C_{20} halves, however, are joined tail to tail so that the two central methyl groups of the molecule are brought into a 1:6 relationship.

The following are the structural formulae of α, β and γ carotene and lycopene:

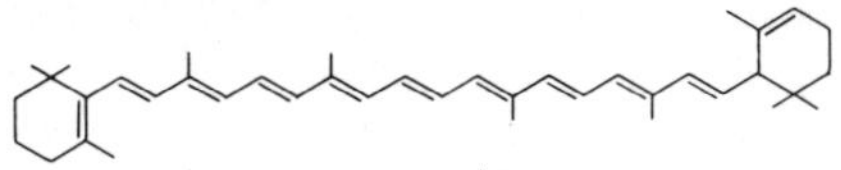

α-carotene

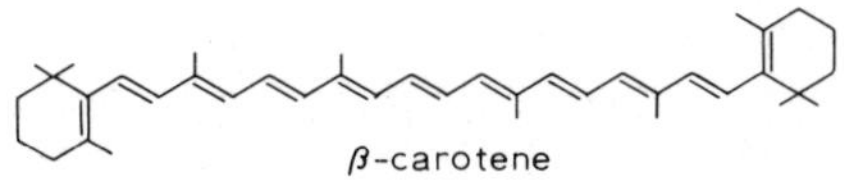

β-carotene

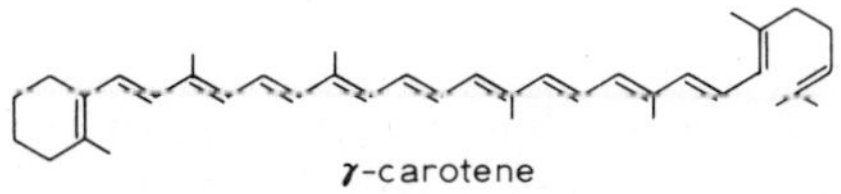

γ-carotene

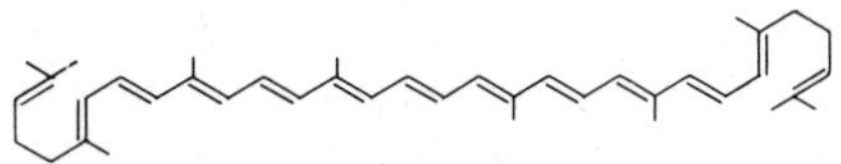

lycopene

Xanthophylls may be considered as oxygenated derivatives of the hydrocarbon carotenoids (carotenes). The structural formula of *lutein* (*β-xanthophyll, cucurbita xanthin*), one of the pigments of egg yolk, is:

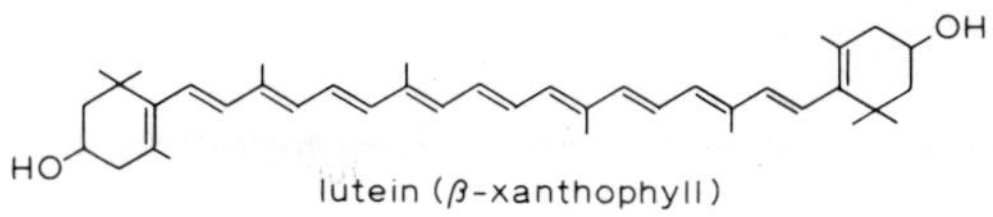

OH

HO

lutein (β-xanthophyll)

Cryptoxanthin, which contains only one hydroxyl group, is another example of a xanthophyll. It is one of the chief pigments in yellow corn, paprika, papaya and mandarin orange.

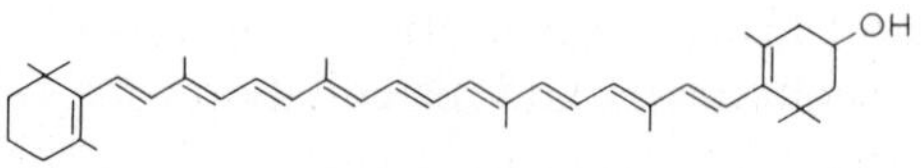

cryptoxanthin

Cryptoxanthin and lutein were shown by Zechmeister and Turzon to be present in the pigments of orange peel (*Citrus aurantium*) in the form of esters. Saponification, followed by fractional extraction of these pigments with petroleum ether and methanol (90%), yielded a carotene fraction (28.1 mg/kg peel) and a xanthophyll fraction (49.8 mg/kg peel). Both fractions were examined chromatographically. The bulk (95%) of the carotene fraction was identified as crystalline cryptoxanthin ($C_{40}H_{56}O$); this fraction also contained β carotene and traces of α-carotene. β-Xanthophyll was found to constitute about a third of the xanthophyll fraction. Curl separated the carotenoid pigments of Valencia orange juice, by counter-current distribution in a petroleum-methanol system. A large number of carotenoids, mostly xanthophylls, were isolated. According to Rother, only 2 to 5% of carotenoid pigments in natural orange juice are hydrocarbon carotenes, the remainder being xanthophylls.

In fruit, xanthophylls occur mainly as esters of fatty acids. Thus the fruit pigment *psysalien* is *zeaxanthin dipalmitate.*

The physical state of the carotenoids in nature varies with the source. In milk, butter and the fat-rich animal tissues, the carotenoids occur in solution in the lipid phase. Similarly, in carrots and most root vegetables, the carotenoids are found in solution in the oil droplets. In leaves, the carotenoids are found in the plastides, in combination with proteins. Protein–carotenoid complexes are widely distributed in nature. The exact character of the bond between the protein and the pigment is not known. Very often the protein is associated with a lipid (lipoprotein) and this latter contains the carotenoid in the adsorbed or dissolved state. However, direct combination of carotenoids to proteins as true prosthetic groups is also known. The color of the complex depends not only on the carotenoid fraction but also on the protein moiety and on the nature of the bond between the two. Thus, colors not usually found in the carotenoids (blue, black) may be observed in such complexes.

12-3. BREAKDOWN OF CAROTENOIDS

Owing to their highly unsaturated nature, carotenoids are apt to oxidize very quickly, particularly at the double bonds. There is some indication that the oxidation and subsequent disintegration of the carotenoids is initiated at the end of the molecule and does not occur at random; the process always takes place at the open end rather than at the terminal ionone ring.

As double bonds are saturated and finally broken down, the characteristic color of carotenoids is bleached. Certain parts of the molecule resist oxidation and appear intact in the mixture of final oxidation products. Thus, the final breakdown of β-carotene results in the formation of ionone, a ketone with the smell of violets:

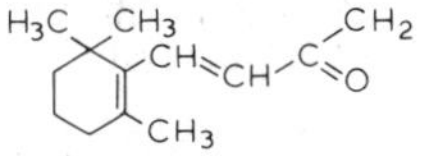

β-ionone

Hay which has been dried in a field in the sun smells of ionone, its total carotene content having been practically destroyed.

Carotenoids are much more stable to oxidation in their natural form than in pure systems. Crystalline pure lycopene or a solution of lycopene in chloroform fades in a matter of a few hours when exposed to air, while the same pigment in its natural form in tomatoes is quite stable.

The oxidation of carotenoids and the autoxidation of fats have many points in common and are often interrelated in food systems. Free radicals formed in the course of fat oxidation may participate in the oxidative attack on carotenoids. The enzyme *lipoxidase*, which is important in the oxidative degradation of fats in grains and vegetables, may also take part in carotenoid oxidation.

The most important single factor in the oxidation of carotenoids is the presence of oxygen or strongly oxidizing reagents. The destruction is more rapid at high temperature. The effect of temperature is complex: it does not only accelerate oxidation directly, but it may also render the carotenoid more susceptible to breakdown by denaturing the protective protein. However, in the absence of air, carotenoids can withstand relatively high temperatures. The effect of moisture is similar to the phenomena observed in fat autoxidation. Bleaching is much more rapid in the absence of water. Moisture content levels above BET monolayer values (see Section 4-7) have a protective effect on carotenoids.

12-4. VITAMIN A

The presence of nutritionally important accessory factors in cod liver oil was demonstrated early in our century. The ability of cod-liver oil to prevent rickets and cure an eye disease known as *xerophthalmia* was assigned to a so-called "fat soluble vitamin A". In 1922 McCollumn reported on investigations which proved that cod-liver oil contains, in fact, two different vitamins, one effective against rickets and the other necessary for the prevention and cure of the eye disease. The name vitamin A was reserved for the latter. Eight years later, mainly as a result of work by Karrer, the structure of vitamin A was elucidated and its relation to carotenoids was established.

A small number of closely related substances show vitamin A activity. The substance first recognized as vitamin A is now termed retinol or vitamin A_1.

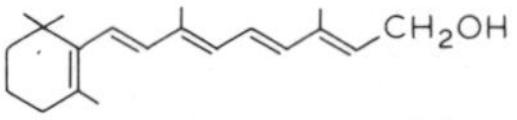

CH_2OH

retinol (vitamin A_1)

The relationship with carotenoids is evident. As far as the carbon chain is concerned, β-carotene consists of two molecules of vitamin A, bound tail to tail. β-Carotene is converted to vitamin A, with the help of enzymes present in the intestinal mucosa of animals, hence the name of *provitamin A* given to β-carotene. The conversion involves oxidation at the middle point, by means of a specific enzyme *β-carotene-15, 15'-oxygenase.* A peroxide is formed. Cleavage of the peroxide yields two molecules of *retinal* (the aldehyde form of vitamin A). The majority of retinal is reduced to retinol by a non-specific enzyme requiring NADH or NADPH. Retinol is carried into the blood stream and any amount in excess of the required level in blood is stored in the liver, in the form of fatty-acid esters. In the blood plasma, retinol is carried by a very specific protein with a molecular weight of 21,000, known as the *retinol binding protein.* Disorders in protein synthesis in the body may cause symptoms of vitamin A deficiency, not because of lack of vitamin A, but rather because of inadequate supply of the retinol binding protein.

Although vitamin A deficiency may cause disorders and lesions in almost every system of the body, and even death, the best known function of this vitamin is connected with vision. In the eye, retinol is oxidized to retinal (formerly called retinene). Retinal combines with certain proteins termed *opsins*, to form the so-called *visual pigments* of the retina. One of these pigments, the visual purple, is also called *rhodopsin.* While the most stable form of retinal is the all-trans isomer, only the 11-cis isomer (*neo-retinene b*) can combine with an opsin. When the visual pigment is exposed to light, the 11-cis isomer is transformed to the all-trans isomer and the retinal-opsin complex is decomposed. The bleaching of the vision pigment by light energy finally creates a pulse which is transmitted to the optical nerve. The mechanism by which all-trans retinal re-isomerizes to the all-cis form in order to combine with opsin is not entirely clear.

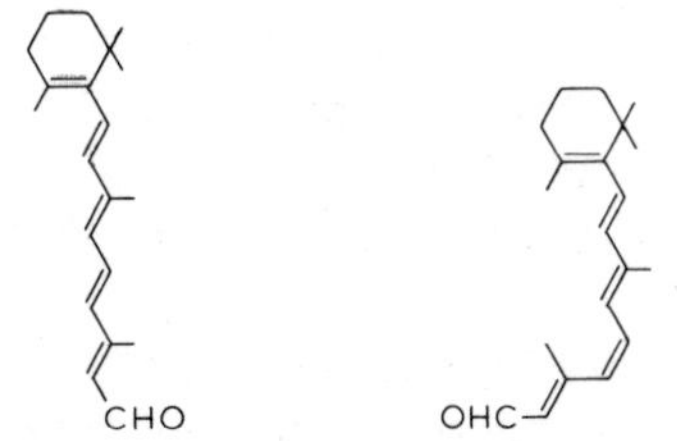

CHO

all-*trans* retinal

OHC

all-*cis* retinal

Very little is known at the present time regarding the *systemic action* of vitamin A in the human body, namely, its action as a growth promoting factor. Although a number of enzyme systems are affected by vitamin A deficiency, vitamin A does not appear to be a co-factor in the true sense.

Vitamin A as such, is not found in plants, yeast, fungi or bacteria, and the herbivorous animal obtains it only from its precursors, the carotenes. Theoretically, one molecule of β-carotene should produce in the animal body two molecules of vitamin A, however, the actual yield is somewhat less than that. Carotenoids with one open ionone ring will procedure only one half the amount of vitamin A, and those with no ionone rings, such as lycopene, will have no vitamin A activity. Liver is an excellent source of vitamin A as such (Hippocrates, ca. 500 BC, knew of night blindness and of its cure by eating liver). Small but not insignificant quantities of vitamin A and its precursors are found in egg yolk and milk and because these substances are soluble in fat, they are found in cream and butter. Synthetic vitamin A is produced commercially and used extensively in the fortification of many foods, notably margarine. The stable commercial form is usually an ester of retinol with acetic acid or palmitic acid.

The international unit (I.U.) accepted for this vitamin is 0.6γ (microgram) of pure β-carotene or 0.3 γ of pure vitamin A-alcohol. Green and yellow vegetables contain between 1500–20,000 I.U. vitamin A per 100 g; butter 1000–4500 I.U. and calf liver between 10,000–160,000 I.U. vitamin A per 100 g.

The recommended requirements for the average adult are about 5000 I.U. from combined sources (of carotene or vitamin A).

Prior to the availability of synthetic vitamin A, the most important natural source of it was the fish liver oils, from which it was recovered by molecular distillation, after appropriate saponification.

The recommended human allowance can be easily provided by one egg a day, plus one liter of milk, 25 g of butter and 100 g of fresh tomatoes. Animals are provided with vitamin A by eating grass containing ample quantities of carotene.

Since both vitamin A and its precursor, β-carotene, are soluble in fats and oils, there is always a danger that when the oils become rancid (due to oxidation) the vitamin will suffer considerable losses. This is true to a large extent of such food products as butter or vitamin-enriched margarines subjected to prolonged storage.

Vitamin A can also be destroyed to some extent by the action of light and, in particular, by the ultraviolet portion of the spectrum. The effect of packaging is therefore an important factor in the retention of vitamin A in products packed in glass, such as milk, or in other transparent forms of packaging.

12-5. FUNCTIONS OF CAROTENOIDS

For a long time carotenoids have been regarded, as have many other secondary products, as plant waste products or, at best, as compounds created by the plant to invite insects for cross-pollination. However, such theories could not explain the invariable presence of carotenoids in chloroplasts alongside chlorophyll. The carotenoids are constantly synthesized, and are specifically incorporated into the structure of the photosynthetic apparatus. It was demonstrated long ago that all the plastid pigments, and not merely chlorophyll, could mediate photosynthetic oxygen evolution and it is now generally accepted that light absorbed by the so-called "*acces-*

sory pigments", the carotenoids, contributes to photosynthesis through the ability of these pigments to transmit the accumulated light energy to the chlorophyll.

Recently, it has been shown that in the absence of carotenoid pigments the photosynthetic apparatus is rapidly destroyed by chlorophyll-catalyzed photooxidation. One of the main biological functions of carotenoids seems to be *photoprotection*, i.e. protection of cells and tissues against harmful effects of light. This effect has been observed not only in green plants but also in microorganisms and even animals and man.

12-6. CAROTENOIDS IN FOOD SYSTEMS

Carotenoids in foods are important mainly as pigments. As explained above some carotenoids are also an important source of vitamin A.

Mixtures of natural carotenoids extracted from plant tissue (e.g. orange peel), and synthetic β-carotene are commercially available as food colorants, the latter both in the oil-soluble and in the water-dispersible forms. However, since in food systems carotenoids are often in chemical combination with other constituents, and since the visual appearance of the food is affected by the nature of such combinations, coloration with added carotenoids is not always easy.

The problem of chemical unstability of carotenoids was discussed in Section 12-3. Loss of carotenoids is not considerable in high moisture systems such as canned and frozen vegetables. It is, however, an extremely severe problem in fat rich systems (butter, margarine) and in low moisture foods (dehydrated vegetables and fruits). The deterioration symptoms are not only a loss of the characteristic color, but also the formation of undesirable odors (hay-like, violet-like, etc.), due to breakdown products. Dehydrated vegetables should be packed in opaque containers, under nitrogen. A method which prevents destruction of carotene in dehydrated carrots by coating the pieces with a thin film of starch before dehydration is being used commercially.

The bleached color of the carotenoids, caused by oxidation, is often a very important indication of the deterioration of a food product. Such is the case, for instance, in the deterioration of citrus essential oils which, in the natural state, are colored by carotenoids but which become bleached upon oxidation.

The characteristic color imparted to foods by carotenoid pigments is not always desirable. Freshly milled wheat flour has a yellowish hue which is unacceptable commercially. The dark color of unrefined edible oils is partly due to the presence of carotenoids. *Bleaching* of flours occurs during storage in air (*ageing*) or can be accelerated by means of strong oxidizing substances. For special purposes, cream is sometimes bleached with peroxides. The pigments of edible oils are removed by adsorption on special clays (see Section 11-3-5).

SELECTED BIBLIOGRAPHY TO CHAPTER 12

Borenstein, B and Bunnell, R.H., Carotenoids: Properties, Occurrence and Utilization in Foods, Adv. in Food Research, **15** (1966), 195-276.

Curl, A.L. and Bailey, G.F., The State and Combination of the Carotenoids of Valencia Orange Juice, Food Research, **20** (1955), 371.

Goodwin, T.W., The Comparative Biochemistry of the Carotenoids, Chapman and Hall, London, 1952.

Isler, O. (Ed.), Carotenoids, Birkhauser Verlag, Basel, 1971.

Isler, O., Ofher, A. and Siemers, G.F., Synthetic Carotenoids for Use as Food Colors, Food Technol., **12** (1958), 520.

Joyce, A., The Effect of Heat Treatment on Some Plant Carotenoids, Proc. Symp. Color in Foods, N.A.S., 1953.

Karrer, P. and Jucker, E., The Carotenoids, Elsevier Publishing Co., Amsterdam, 1950.

Mackinney, G. and Chichester, C.D., The Color Problem in Foods, Adv. In Food Research, 5 (1954), 301.

Zechmeister, L. and Deuel, H.J., Jr., Stereochemical Configuration and Provitamin A Activity, Arch. Biochem. Biophys., **36** (1952), 81.

CHAPTER 13

TERPENES, ESSENTIAL OILS

Ein kostlicher, teuer Schatz

CONRAD GESNER, 1555

13-1. OCCURRENCE

The *essential oils*, those odoriferous substances found in practically every plant, are very numerous and widely distributed in many different parts of the same plant: in the roots, stem, leaves, flowers and fruits.

Here are a few examples of such essential oils listed according to their source:

In roots—licorice, caradamum, asparagus oil.
In stem—turpentine.
In leaves—mint, geranium oil, eucalyptus oil.
In flowers—jasmin, neroli oil, rose oil.
In fruits—orange and lemon oils, bergamot oil, essences of strawberries, grapes, etc.

It is interesting to note that, when one plant has essential oils in several parts of it, they always differ in their constitution; thus, for instance, an orange tree has essential oils in its flowers (oil of neroli), in its young leaves and twigs (oil of petitgrain), in the yellow peel (flavedo), in its fruits (oil of orange) and, finally, a specific aroma in the orange juice—all of these are quite different in odor and in their chemical composition.

In some fruits the aromatic material may be actually dissolved in the juice, however, in many cases in fruits and in the leaves, the essential oils are secreted in numerous tiny oil sacs or glands located in the epicarp adjacent to the chromoplasts. These glands or canal-like intercellular receptacles are ductless and have no walls of the usual type in the cells, but are bounded by the debris of degraded tissues.

The physiological function of essential oils in plant metabolism is still obscure. While the odoriferous principle in the leaves or flowers can be assumed to be useful in attracting insects to the pollen, no such property can be ascribed to the oil present in other parts of the plant. The oil may possibly act as a protection against insect attack. For the moment, however, there is no definite proof for either theory, and the only alternative that remains is to regard essential oils, as many other secondary products, anthocyanins, alkaloids, tannins, etc., as waste products of plant metabolism.

Essential oils are mixtures of various volatile organic substances along with some non-volatile waxy materials. The term "oil" in their name does not refer to any chemical characteristic but rather implies that these substances are insoluble in water but soluble in non-polar solvent.

Chemically, the greater part of most essential oils consists of terpenoids and their derivatives. *Terpenoids* are naturally occurring isoprenoid hydrocarbons (terpenes) and their oxygenated derivatives. By this definition, carotenoids would belong to this group. In fact, many texts classify carotenoids as a special group of terpenoid substances.

13-2. STRUCTURE AND NOMENCLATURE

Terpenes may be classified according to the number of isoprene units in their molecule, as follows:

Monoterpenes	2 isoprenes	C_{10}
Sesquiterpenes	3 ,,	C_{15}
Diterpenes	4 ,,	C_{20}
Sesterterpenes	5 ,,	C_{25}
Triterpenes	6 ,,	C_{30}

By this classification, vitamin A is a diterpene and carotenoids, which all have 40 carbon atoms, would be tetraterpenes. *Rubber*, a high polymer with an isoprenoid structure, would be a polyterpene.

In mono-, sesqui-, di- and sesterterpenes, the isoprene units are linked head-to-tail. In triterpenes, two sesquiterpene structures are linked tail to tail. The chain may be open (acyclic terpenes) or closed to form one or several rings (cyclic terpenes).

The names by which we designate the individual terpenoids are suggestive of their most common natural source and they carry a suffix designating the principal chemical function of the molecule. Systematic names are not used. Thus, ment*hane* is a saturated alkane, limon*ene* is an alkene (has double bonds), ment*hol* is an alcohol, carv*one* a ketone and citr*al* an aldehyde.

13-3. MONOTERPENES

Many essential oils found in nature are mixtures of a number of monoterpene hydrocarbons of the general formula, $C_{10}H_{16}$, hydrocarbon sesquiterpenes ($C_{15}H_{24}$), both of which serve primarily as carriers for the more important classes of oxygenated compounds (alcohols, aldehydes, ketones, acids and esters) which, although present in much smaller quantities, are usually the bearers of the characteristic odor of the oil in question. All mono- as well as sesquiterpenes are volatile when distilled by steam.

A few of the simplest monoterpenes, such as *myrcene* (from oil of bay) and *ocimene* have acyclic aliphatic structures and contain three double bonds:

β-myrcene   *cis*-α-ocimene

The bulk of the monoterpene hydrocarbons, however, have a ring formation with two double bonds and may be grouped under the general name of *menthadienes*, from the basic cyclic saturated from *menthane*:

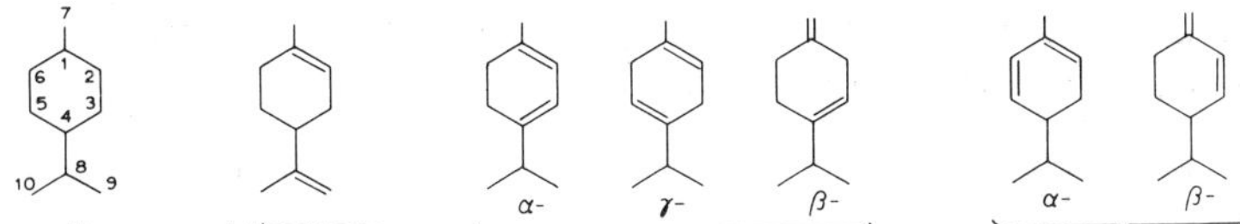

menthane   d-limorene   α- γ- β- terpinene   α- β- phellandrene

d-Limonene constitutes over 90% of orange oil, and its levoratatory modification occurs in Finland turpentine oil, American peppermint oil, etc. *γ-Terpinene* and *α-phellandrene*, which are easily oxidized into a ketone, are found in lemon oil.

The above are strikingly similar in structure except for the position of the double bonds. The following are some monoterpene hydrocarbons with different structures:

CH2

α-pinene β-pinene camphene

sylvestrene d-carene α-thujene

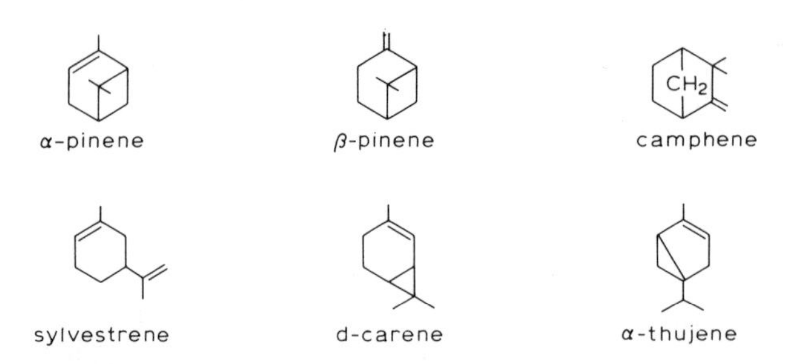

Pinene exists in two optically active modifications. Its α-isomer is the principal constituent of turpentine and is found also in neroli oil.

Examples of the cyclic oxygenated hydrocarbons are:

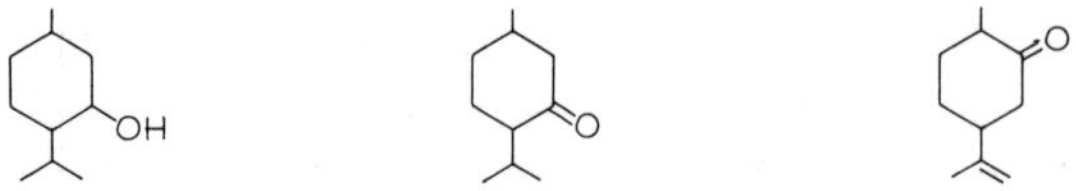

OH
O
O

menthol menthone carvone

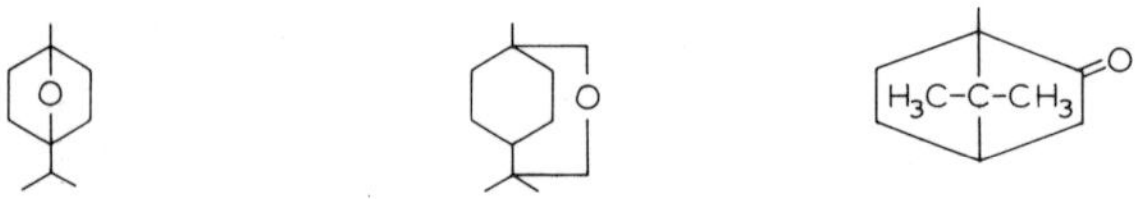

O
O
H3C-C-CH3
O

cineole-1,4 cineole-1,8 d-camphor

Menthol is the main constituent of mint oil, *carvone* is found in caraway and *d-camphor* in Cinnamomum camphore. The two *cineoles* and the *carvone* are products of the oxidation of some of the cyclic hydrocarbons, as will be shown later.

The principal bearers of the specific odors of the essential oil are acyclic oxygenated compounds, mostly alcohols and aldehydes.

The most important alcohols are:

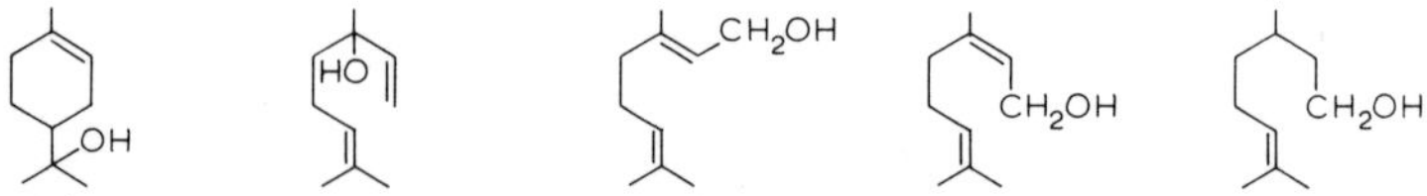

OH
HO
CH2OH
CH2OH
CH2OH

α-terpineol linalool geraniol nerol citronellol

d-Linalool occurs in sweet oranges, and its acetate ester is very prominent in bergamot and other oils. *Geraniol* and *nerol* are merely trans and cis transfigurations of the same alcohol, the first being the main constituent of geranium oil and the second that of the oil of orange blossoms (neroli). Both geraniol and nerol undergo ring closure on acid treatment to yield 1,8 terpine hydrate. The rate of this reaction indicates whether the substance in question is geraniol or nerol: a cis-configuration (nerol) will close more rapidly:

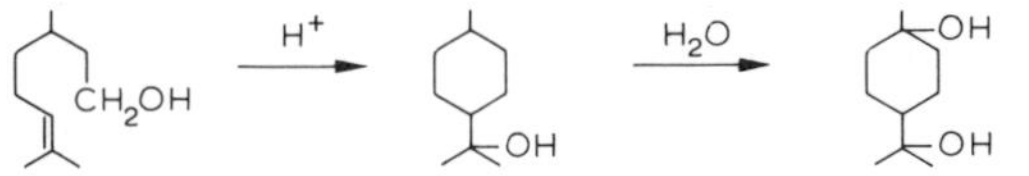

H^+ → H_2O →

nerol α-terpineol 1-8-terpine hydrate

Citronellol is a widely distributed alcohol, which has the odor of roses and is almost always accompanied by geraniol. Since it has an asymmetric carbon atom, it exists in two optically active modifications. While *d*-citronellol occurs in lemon oil, its levorotatory isomer is found in rose oil, both enantiomorphic forms being found in geranium oil.

α-Terpineol is a solid substance, which has a closed ring structure and is optically active. Terpineol can be prepared from limonene by hydrobromination and substitution of the Br by an hydroxyl group:

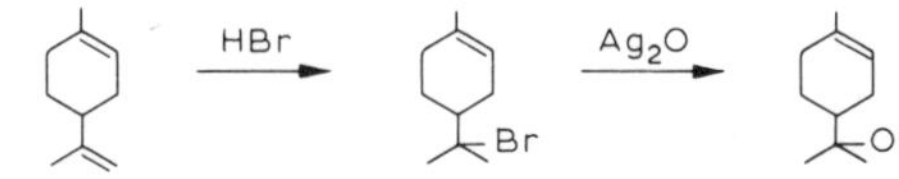

HBr → Ag_2O →

d-limonene α-terpineol

Commercially α-terpineol is prepared from α-pinene by the action of H_2SO_4 in acetic acid. Still more important are the terpene aldehydes which are very widely distributed in essential oils and are greatly valued because they possess characteristic odors and flavors. In fact, the aldehyde value of most of the essential oils is the major determination made when appraising their quality. The most recognized aldehydes are α- and *β-citrals*, and d- and l-*citronellals*:

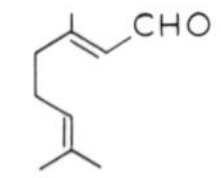

α-citral 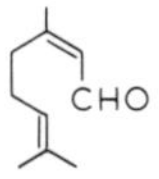 β-citral 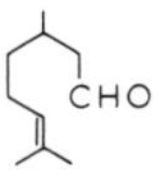 citronellal

methyl-anthranalic acid  methyl-anthranilic acid methyl ester

Citral is widely distributed in nature; it is an important constituent of many essential oils, particularly lemon oil and lemon grass oil. In all these, citral appears as a mixture of the two stereoisomers, α-citral (trans) and β-citral (cis). Citral can be prepared synthetically from methyl heptenone and ethyl chloroacetate using Zn as a catalyst. If citral is condensed with acetone in the presence of weak alkalis, pseudoionone is formed, which, when heated with dilute H_2SO_4, gives a mixture of α- and β-ionone, a synthetic perfume with the odor of violets.

Citronellal, another aliphatic unsaturated aldehyde, has one double bond and, since it has one asymmetric carbon atom, appears in two stereoisomeric configurations, d- and l-.

Methyl anthranilic acid and its methyl ester (*oil* of *wintergreen*), although not terpenoids are of extreme importance as odoriferous substances and occur as the different flavors of certain fruits, such as Concord grapes.

Monoterpenes boil over a range of 170-200°C at atmospheric pressure.

13-4. SESQUITERPENES

The higher boiling fractions of many essential oils, those boiling over the range of 250° and 280°C at atmospheric pressure, consists of sesquiterpene hydrocarbons with the general formula $C_{15}H_{24}$. They are widely distributed in nature but have been investigated only to a limited extent. The sesquiterpenes have a density of approximately 0.90, their viscosity is higher than that of monoterpenes and they have only a faint odor. The structures of a few sesquiterpenoids are shown below:

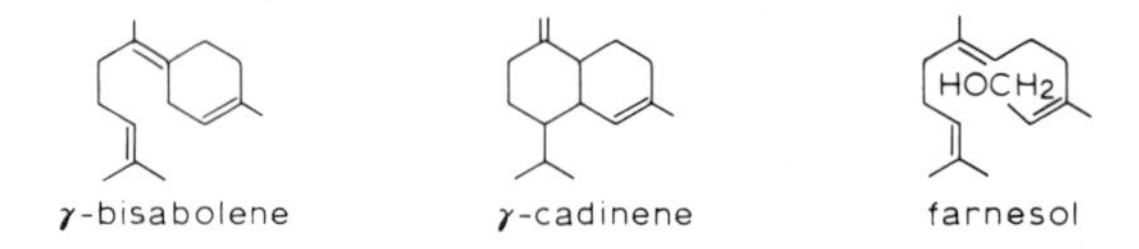

γ-Bisabolene, is found in bergamot oil, and *cadinene*, which is its cyclic isomer (with two rings) is found in the oil of guayule and other plants. The sesquiterpenoid, *farnesol*, which is an alcohol, appears to be widely distributed and is found in small amounts in many essential oils and is considered to be an important intermediate in sesquiterpenoid biosynthesis.

Celery oil contains another sesquiterpene, *β-selinene*, and eucalyptus oil — *α-eudesmol* — both of them having curious isoprenoid structures:

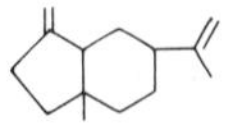

β-selinene 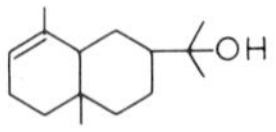 α-eudesmol

In the high boiling fraction of neroli oil, an aliphatic sesquiterpene alcohol is found. It possesses only a faint odor but is capable of fixing other perfumes:

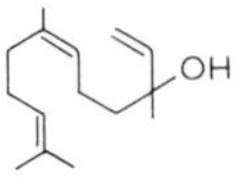

nerolidol

An interesting sesquiterpenoid, *nootkatone*, was isolated by MacLeod from grapefruit essential oil. Nootkatone seems to be responsible for the characteristic, resin-like tone of grapefruit flavor distinguishing this fruit from most other citrus species.

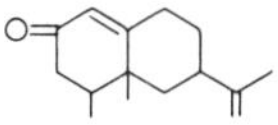

nootkatone

13-5. HIGHER TERPENOIDS

Being non-volatile, higher terpenoids are not odorous components of essential oils. They constitute the bulk of the non-volatile fraction of *oleoresins*; oleoresins are ether extracts of resinous plant tissues. Oleoresins of spices and various vegetables (pepper, tomato seeds, etc.) are valuable flavor concentrates. Many oleoresins have found various non-food applications in pharmacy and industry. Distillation of oleoresins usually yields a volatile essence consisting of lower terpenoids and a residue of higher terpenoids. The group also comprises a considerable number of biologically important substances.

13-5-1. *Diterpenoids. Diterpenoids* are widely distributed. Although the number of theoretically possible steric configurations is very large, only a few basic shapes are found in nature. One group of considerable importance includes the so-called *resin acids*. These are constituents of plant resins. The structure of *abietic acid* (*sylvic acid*), obtained from colophonium (rosin, pine resin) is shown below:

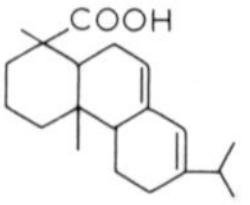

abietic acid

The plant growth hormones *gibberellins* are also diterpenoids. Gibberellins are secreted by phytopathogenic fungi and have first been isolated from the mold *Gibberella fujikuroi*, which is responsible for the "Bakanae disease" in rice. Since then, various gibberellins have been isolated from higher plants as well. Gibberellins are produced commercially and used as plant growth promoting agents in agriculture and as additives in the malting of barley. The basic structure of gibberellins is shown below.

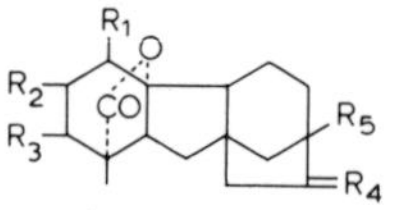

basic structure of gibberellins

Closely related to diterpenes is the isoprenoid alcohol phytol, an essential part of the chlorophyll molecule.

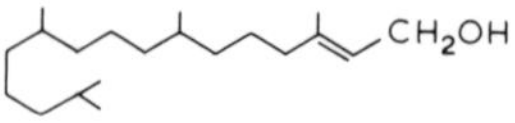

phytol

13-5-2. *Sesterterpenoids.* The natural sesterterpenoids known today are relatively few. Some are fungal phytotoxins. The protective waxes of some insects contain sesterterpenes.

13-5-3. *Triterpenoids.* Triterpenoids constitute the connecting links between terpenes and another class of important fat soluble substances: the steroids (Chapter 14). The isoprenoid hydrocarbon squalene $C_{30}H_{50}$, is a precursor in the biosynthesis of both triterpenoids and steroids. It is found in small quantities in several oilseeds, rice bran and wheat gum, and in considerable concentration in shark liver oil.

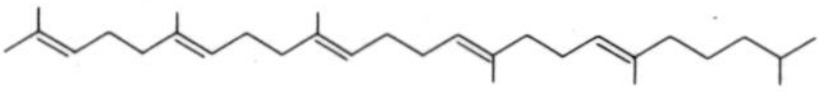

squalene

A group of triterpenoids, the so-called *limonoids*, are bitter substances found in the seeds of certain citrus fruits. These substances are of considerable interest in citrus technology. It is known that the juice of Navel oranges, although sweet when first pressed, becomes extremely bitter if it is allowed to stand in contact with pulp. The bitterness develops rapidly upon heating, e.g. as a result of pasteurization. For this reason, Navel oranges cannot be utilized as a commercial source of processed juice. The same type of bitterness is also found in other varieties of sweet oranges, especially in the early stages of maturation. The bitter principle of such juices was isolated by Higby and identified as *limonin*, a substance closely related to the seed limonoids. Since then a voluminous literature has accumulated on the structure and biochemistry of limonin and its precursors.

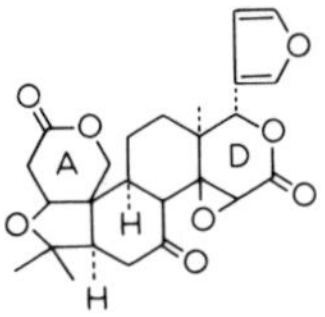

Limonin has two lactone rings designated as the A-ring and D-ring lactones. The development of bitterness has been explained by Maeir and Beverly. According to these authors, limonin is formed from a non-bitter precursor, limonin monolactone, which differs from limonin by the absence of the D-ring lactone ring. The monolactone precursor, later recognized as *limonic acid A ring lactone*, is stable in the intact tissue of the fruit. When the orange is pressed and some of the tissue disintegrates as pulp, rapid lactonization occurs. The existence of an enzyme system which catalyzes this reaction has been suggested.

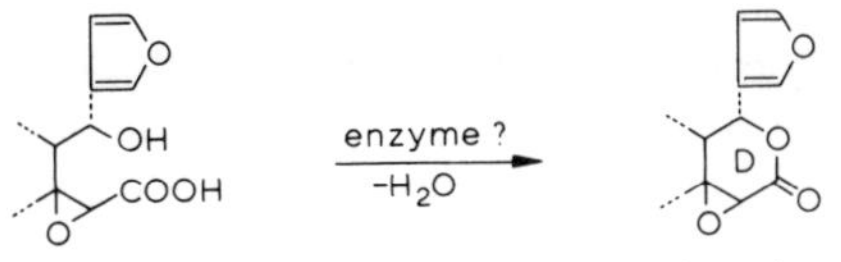

limonin monolactone precursor (not bitter) —(enzyme ?, $-H_2O$)→ limonin (bitter)

Bitterness due to limonin may be prevented or minimized by:

(a) Avoiding the inclusion of too much pulp and "peel juice" containing the precursor (milder pressing).

(b) Separation of the pulp particles immediately after pressing, or

(c) Removal of limonin and its precursor by adsorption on polyamide resins (nylon).

Limonin and its precursor undergo autoxidative degradation. Flavian and Levi have explained the disappearance of bitterness with maturation as a consequence of enzymic degradation of the limonin monolactone.

13-6. CHEMISTRY OF FOOD ODORS

The study of food flavors has been one of the most active fields of food research in the last twenty years. Perhaps the most important parameter of food quality, *flavor*, is a complex sensation arising from the simultaneous perception of odor and taste. Taste and odor are usually classified as the "*chemical senses*" in contrast to the perception of color or consistency which appeal to our "*physical senses*", vision and touch. Despite considerable progress in the recent years, the mechanism of taste and odor perception is one of the most obscure chapters of human physiology.

Odors are sensed when molecules of volatile substances reach the olfactory receptors at the top of the nasal cavity. The relationship between chemical composition and odor is not known. A number of theories exist, but none seems to be entirely satisfactory. Recently, a theory which relates odor to the shape and size of the volatile particles (distinct molecules or associated clusters) has been been receiving considerable experimental support.

The chemical nature of food odors has been lately an extremely productive subject of research. Due to lack of appropriate analytical tools, the characterization of food flavors was, for a long time, based on analogy. Amyl acetate, a simple chemical substance which can be easily synthesized, has an odor very similar to that of bananas; so, the factor responsible for the odor of bananas was assumed to be amyl acetate. This approach, although very useful to the chemist dealing with the development of synthetic flavors, did not provide a sound basis for the chemical definition of natural food odors. The accurate and complete quantitative analysis of food volatiles was made possible by the development of gas chromatography and mass spectrography. With the help of these tools it was soon realized that the odor of foods is seldom due to one or few chemical substances. Head space analysis of orange juice (i.e. gas chromatographic analysis of the atmosphere above orange juice in a closed container) reveals the presence of close to 200 different chemical species. Some of these volatile components are probably unimportant as far as orange juice odor is concerned. However, it is impossible to assign the odor of orange juice to just a small number of components. Many substances, although not major carriers of characteristic orange juice odor, contribute their own "tone", without which the odor would be very different. The same situation prevails in coffee, most fruits and, of course, in wines. The analytical tools mentioned above have enabled the flavor chemist to follow the changes in food volatiles as a result of processing and storage, or to evaluate quantitatively the loss of flavors by evaporation, in concentration or dehydration processes.

13-7. TRANSFORMATION OF TERPENOIDS IN ESSENTIAL OILS

Most of the aliphatic monoterpenes can readily close their ring to become cyclic terpenes, as shown above, during the cyclization of geraniol into terpineol. However, such reactions are not reversible. The following is an example of such transformations in peppermint and spearmint oils.

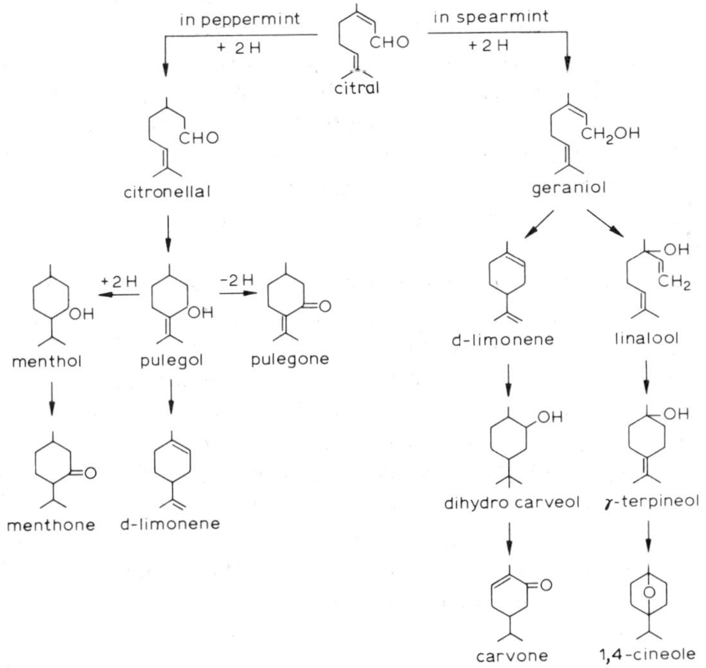

in peppermint
+ 2 H
CHO
citral
in spearmint
+ 2 H
CHO
citronellal
CH_2OH
geraniol
+2 H
-2 H
OH
OH
=O
menthol
pulegol
pulegone
OH
CH_2
d-limonene
linalool
=O
menthone
d-limonene
OH
OH
dihydro carveol
γ-terpineol
O
O
carvone
1,4-cineole

Another type of transformation in essential oil constituents occurs when these are distributed in an aqueous solution. Such is the case, for instance, when d-limonene, which constitutes the bulk of orange oil, is present in orange juice. In such case the essential oil will undergo an acid catalyzed hydration–dehydration, with a resulting off-flavor. This has also been attributed to the conversion of citral to *p*-cymene as well as to the oxidation of d-limonene to carvone and carveole.

13-8. TECHNOLOGY OF ESSENTIAL OILS

13-8-1. *Extraction of essential oils.* In many cases the crude essential oils are still extracted by steam distillation, mostly with primitive equipment in their native places of origin. The plant materials, such as menth, geranium, rose, etc., are distilled with water usually in copper vessels and the distillate is collected from the condensers, the oil phase being subsequently separated from the water and filtered off. When the essential oil in question is liable to suffer from high temperatures, a vacuum is applied.

However, many essential oils do not lend themselves to steam distillation and in such cases methods using various organic solvents are used. The extraction of essential oils from flowers, such as jasmin and cassia, are usually performed by means of diffusion, percolation or maceration with solvents such as petroleum ether, the solvent being subsequently distilled off in vacuo at a low temperature. The solvents will in these cases naturally extract a number of other substances in addition to the essential oils, e.g. lipids, sterols, pigments, etc., and very often the resulting products are quite solid (called "*concretes*").

Another process used for the extraction of essential oils from flowers is *enfleurage*, which consists in spreading the flowers upon a neutral, purified fat which slowly absorbs the essential oil. This method of preparation is used in various cream pomades and lotions.

A number of essential oils used in flavoring food products, especially citrus fruit oils, are so delicate that none of the above methods can be successfully used. These oils are pressed by hand or by various, so-called "cold-press" methods making use of the fact that the essential oils secreted in the glands or sacs of the plant tissue are maintained under a certain turgor pressure. When the glands are ruptured mechanically, the oil will eject with considerable force to a relatively great distance. The extraction of the essential oils in these cases consists, usually, in piercing or rupturing the oil glands and collecting the oils in the form of water emulsions, which are then separated by centrifuging. In modern citrus juice extractors, juice and the peel oil are pressed simultaneously in one operation.

13-8-2. *Terpeneless oils.* As has been pointed out earlier most essential oils are mixtures of oxygenated compounds, which are the principal odoriferous agents, with terpenes and sesquiterpenes, which are hydrocarbons but which have no aroma of their own. Moreover, these terpenes differ greatly from the oxygenated compounds in their solubility, being very poorly soluble in weak alcohols; they also oxidize into oxides and peroxides and are apt to polymerize. It has, therefore, always been the aim of the manufacturer of flavoring extracts to recover the valuable portion of the essential oils, i.e. the oxygenated compounds, in some sort of "concentrated form". This process, consisting primarily in the elimination of the less odorous terpenes, is called *deterpenation* and the resulting oxygenated compounds—*terpeneless oils.*

There are three main methods for the deterpenation of essential oils: (1) By fractional distillation in vacuo; (2) by counter-current fractional distribution using different solvents; and (3) by chromatographic separation in columns containing silica gel or other adsorbing material. The concentration ratio achieved by deterpenation is expressed as "*fold*". Thus "7-fold lemon oil" refers to lemon essential oil which has been concentrated by removing six parts of terpenes from seven parts of original oil.

13-8-3. *Aroma recovery.* Many operations, applied in food processing, cause considerable losses in flavor and especially in aroma. Such lossses are caused in the first place during the blanching of vegetables and fruits. While blanching by steam or scalding by hot water are mainly required in order to inactivate the enzymes, which might otherwise have a deleterious effect on the foods during subsequent operations and storage, at the same time blanching may cause substantial losses in aroma due simply to the fact that the odoriferous substances are distilled off during this process.

Even more important are the losses in aroma sustained during concentration and dehydration, for no matter how low the temperature maintained during these processes of evaporating water, the volatile constituents will always tend to be removed first and be lost from the food. This is the major reason why most concentrated juices and dehydrated fruits and vegetables retain very little of their fresh odor and flavor.

One of the methods applied in the food industry to overcome evaporative loss of flavor is known as *aroma recovery* (essence recovery). It is applied mainly to evaporative concentration processes. The flavor volatiles are recovered from the vapors in the form of highly concentrated (several hundred fold) *essences*, which are added back to the concentrate. This process is widely applied in the manufacture of fruit juice concentrates and soluble coffee. It is interesting to note that, generally, the chemical stability of the flavor compounds in the recovered essence is not as good as in the original juice, probably due to a change in the physical state of dispersion.

SELECTED BIBLIOGRAPHY TO CHAPTER 13

Braverman, J.B.S., Citrus Products—Chemical Composition and Chemical Technology, Interscience Publishers, New York, 1949.

Gunther, E., The Essential Oils, 6 Vols., Van Nostrand Co., New York, 1948-1952.

Flavian, S. and Levi, A., A study of the natural disappearance of the limonin monolactone in the peel of Shamuti orange, J. Food Technol., 5 (1970), 193.

Maier, V.P. and Beverly, G.P., Limonin Monolactone, the nonbitter precursor responsible for delayed bitterness in certain citrus juices, J. Food Sci., 33 (1968), 488.

Merory, J., Food Flavorings, Avi Publishing Co., Westport, Conn., 1960.

Newman, A.A. (Ed.), Chemistry of Terpenes and Terpenoids, Academic Press, New York, 1972.

CHAPTER 14

STEROIDS AND TOCOPHEROLS

Besides the carotenoids and terpenes, the isoprenoid structure is found in two additional groups of naturally occurring substances: the *steroids* and the *tocopherols*. Members of these groups are of considerable importance in nutrition, food toxicology and physiology.

14-1. STEROIDS

14-1-1. *Structure and occurrence.* The term "steroid" is derived from *cholesterol*, the name of a white crystalline substance isolated from gallstones and animal fats early in the 19th century (cholesterin or cholesterol from the Greek *chole* = bile and *stereos* = solid).

The characteristic feature of steroids is a carbon skeleton consisting of three cyclohexane and one cyclopentane rings, fused in the following manner:

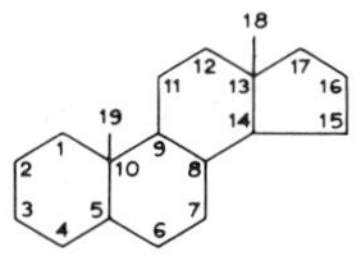

One class of steroids, the *sterols*, are particularly abundant. Sterols contain a hydroxyl group at C_3 and an aliphatic radical attached to C_{17}.

Among steroids of importance in foods, nutrition, toxicology, biotechnology and physiology, one can cite the D vitamins, the saponins, steroidal alkaloids, cholesterol, the steroid hormones and the bile acids.

Steroids are fat soluble substances. They are found in animal and plant tissue mostly in solution in a fatty phase or in dispersion as micelles.

Various steroids are found in the unsaponifiable fraction of vegetal oils and animal fats. A certain proportion of the steroids is removed by steam distillation in the process of deodorization of edible oils (see Chapter 11).

14-1-2. *Vitamin D.* As early as 1838 *rickets*, a children's disease, called in some places "the English disease", was ascribed in France by Jules Guerin as due to faulty diet. In 1906 Hopkins postulated that the absence of an "accessory foodstuff" might cause rickets. However, until about 1922, most of the prominent pediatricians in Europe and America were skeptical about the value of cod-liver oil in curing rickets.

In the same period, it was noted that exposure of rachitic animals to sunlight resulted in recovery. Ultraviolet radiations were found to enhance the antirachitic potency of foods given to experimental animals. From these observations, it was concluded that the food contained provitamins which were converted to the vitamin upon irradiation. Further studies showed that these precursors were sterols present in the non-saponifiable factor of the fats. The antirachitic factor was named *vitamin D* and its precursors pro-vitamins D.

Several facts emerged from mere observation and accumulation of data: the strata of population which suffered most from this disease were the poor sections of town dwellers whose children had very little opportunity to enjoy sunshine. But, even in sunshine-rich countries, rickets had been reported among the classes who dwelt with their children in slums away from the sun. In contrast to this, the Eskimos, who spend their childhood mostly in dark huts during the long, arctic night, are free from rickets.

Today, the explanation for all these peculiar phenomena is simple: the Eskimo's diet is rich in fish-liver oils which contain sufficient anti-rickets vitamin; on the other hand, children who live in countries with abundant sunshine are exposed to the ultraviolet rays of the sun which are able to synthesize the anti-rickets vitamin in the children's own bodies.

Deficiency in anti-rickets vitamin causes insufficient deposition of mineral matter, calcium and phosphates, in the bones, which then become soft and apt to bend. The blood of a rickets patient is poor in these minerals and high in the enzyme *alkaline phosphatase.* The teeth are also affected. Vitamin D deficiency in adults is called *osteomalacia.*

It has been established that *ergosterol,* found in yeast and fungi, and 7-dehydrocholesterol found in various fats and oils, are precursors of the anti-rickets vitamins. These two substances (and a few other sterols to a lesser extent) acquire antirachitic properties on irradiation by ultraviolet rays and the products so obtained are called:

Vitamin D_2 or *ergocalciferol*—by irradiation of ergosterol, and
Vitamin D_3 or *cholecalciferol*—by irradiation of 7-dehydrocholesterol.

The following are the formulae of vitamins D_2 and D_3 and their precursors:

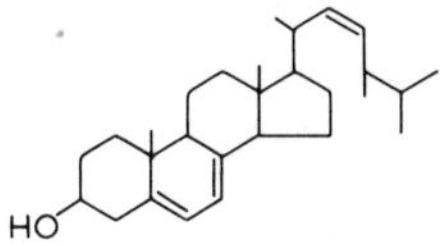

HO

provitamin D_2 (ergosterol)

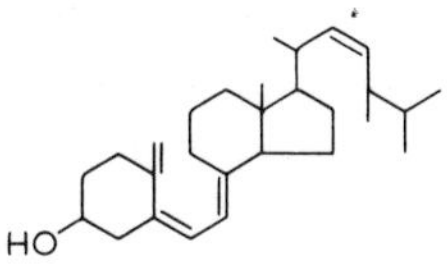

HO

vitamin D_2 (ergocalciferol)

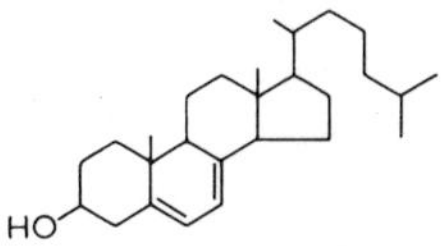

HO

provitamin D_3
(7-dehydrocholesterol)

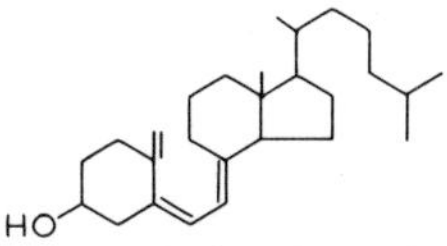

HO

vitamin D_3 (chlolecalciferol)

When the provitamins are irradiated, the ring B opens at position 9, 10.

Several other provitamins have been found in nature, all of them being sterols in which a hydroxyl is attached to carbon atom 3 and with a 5,7-dienic group.

Vitamin D_1, the original vitamin D, was later shown to be a mixture of vitamin D_2 and *lumisterol*, a stereoisomer of ergosterol formed in the conversion of this provitamin to vitamin D_2.

Vitamin D enhances the absorption of both calcium and phosphorus from the intestines and increases resorption of phosphate in the kidney. It is therefore essential for the normal deposition of calcium phosphate in bones. Its mechanism of action is believed to be associated with the synthesis of special proteins responsible for the transport of calcium.

The accepted international unit of anti-rachitic vitamins is 0.025 microgram of pure crystalline vitamin D_3 and the preventive dose is considered to be 400 I.U. per day.

In contrast to all other vitamins, only very few foods contain the

D vitamins. The richest sources are fish-liver oils (cod, 10,000 units per 100 grams, halibut 140,000), fish (mackerel 1100 units, sardines 1400) and egg yolk (265 units per 100 grams). Most of the vegetable oils and animal fats, and consequently, also milk, have only slight activity. Some fatty food product such as margarine are usually enriched artificially with vitamin D.

14-1-3. *Saponins and steroid alkaloids. Saponins* are plant glycosides that form frothing, soapy solutions in water. They are divided into two groups following the nature of the aglycone: triterpenoid saponins and steroid saponins. They are known to cause hemolysis, i.e. dissolution of the red cells of the blood. Plant extracts containing saponins have been used by Indian tribes as fish poisons. Similar extracts have been used in the manufacture of several popular foods as heat-resistant foaming agents. In consideration of their potential toxicity, the use of saponins and saponin-containing extracts has been prohibited in many countries. Glycyrrhizin is a saponin found in the licorice root. It is fifty times sweeter than sucrose and has been considered as an artificial sweetener. Its structure is shown below:

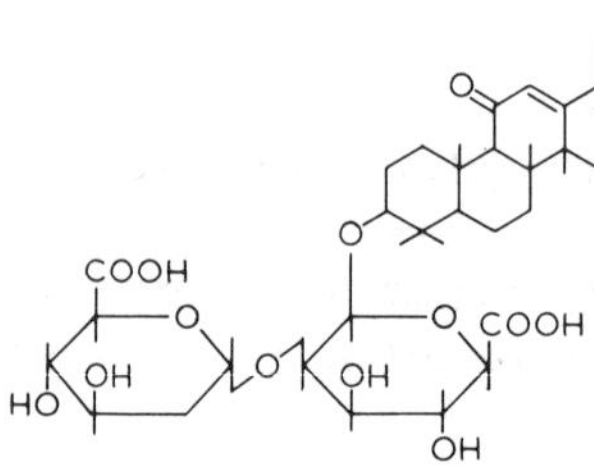

COOH O O COOH OH OH HO OH

glycyrrhizin

Glycyrrhizin, which is a triterpenoid saponin, is commercially available as the ammonium salt.

Steroid saponins are less abundant but of considerable interest in toxicology and pharmacology. They include the "cardial glycosides" of digitalis: *digitonin*, *gitonin* and *tigonin*.

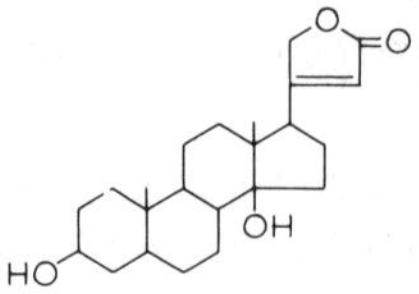

O O OH HO

digitoxigenin
(aglycone of digitonin)

Digitonin reacts with sterols such as cholesterol, to give insoluble addition compounds, and therefore serves as a laboratory reagent for the isolation and determination of sterols. Its reaction with cholesterol and the resulting disorganization of the cell membrane may be the cause of its hemolitic effect.

Solanine, a mildly toxic substance found in potatoes, is both a glycoside and an alkaloid containing a tertiary nitrogen. It is present in the sprouts and immature tubers of potato. A similar substance, *tomatine*, occurs in unripe tomatoes, but disappears with maturation.

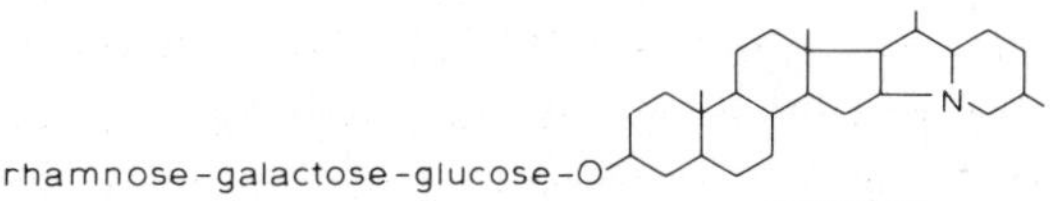

N

rhamnose-galactose-glucose-O

solanine

14-1-4. *Cholesterol.* Cholesterol is the most abundant sterol of the animal kingdom. It occurs as a structural element of cell membranes of many tissues in conjunction with phospholipids. Its function may be associated with its property to impart to lipid phases considerable water-adsorbing power. Its relation to blood vessel diseases has attracted much popular attention. Its dietetic aspects (cholesterol content of foods, effect of intake of particular foods on the cholesterol level in blood) are of importance in food formulation.

The structural resemblance between steroids (e.g. cholesterol) and squalene (see Chapter 13) is interesting. Squalene is indeed a precursor in the biosynthesis of the steroid skeleton.

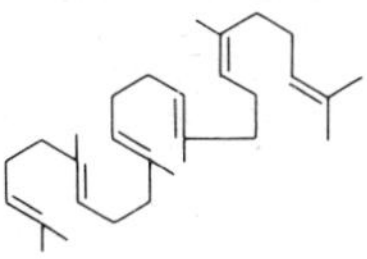

HO

squalene

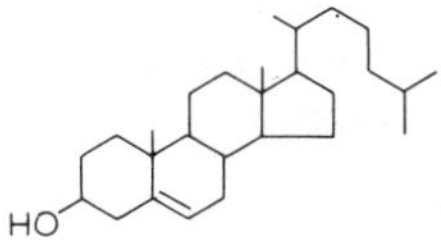

cholesterol

One type of gallstones consists mainly of crystals of cholesterol. This is, in fact, the source from which cholesterol was first isolated. A much better known pathological condition connected with cholesterol is the deposition of this substance in vascular walls (*atherosclerosis*).

As most other sterols, cholesterol also occurs in ester combination with fatty acids.

Closely related to cholesterol is *lanosterol*, found in the fatty component of wool (wool grease = lanolin). Lanosterol is an effective fat–water binding agent, hence the use of lanolin in "moisturizing creams" and other dermatological preparations. Lanosterol is an intermediate in the biosynthesis of cholesterol.

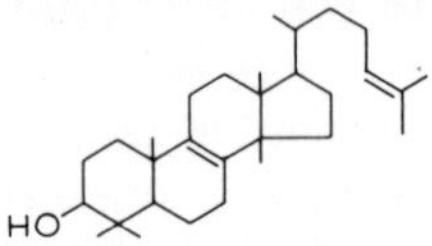

HO

lanosterol

In addition to its functions as a structural element in the cell membrane, cholesterol serves as a precursor in the biosynthesis of ergosterol (hence vitamin D), steroid hormones (see below) and *bile acids.* The latter are surface active substances secreted into the intestine by the gall bladder. The emulsify the fats in the intestinal tract and thus facilitate their digestion and absorption. The bile acids are synthesized in the liver, from cholesterol. Over 90% of the quantity secreted is reabsorbed and recycled. The structure of one of the many bile acids, *cholic acid*, is shown below. The emulsifying power is evident from the relative abundance of hydrophilic groups, especially the dissociable carboxyl.

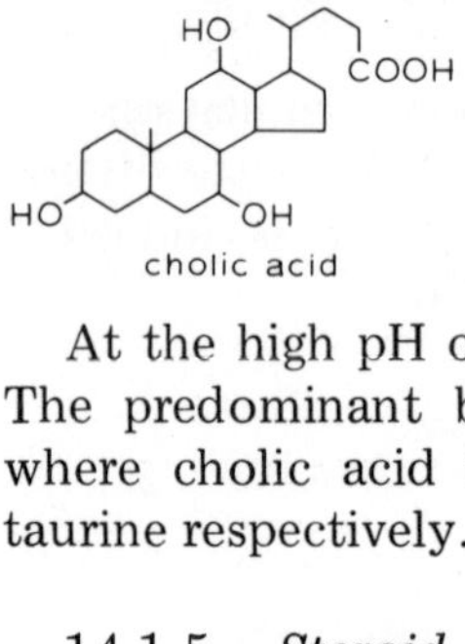

HO COOH HO OH

cholic acid

At the high pH of bile, the acids are present in the form of salts. The predominant bile acids are *glycocholic* and *taurocholic acids*, where cholic acid is combined by a peptide bond to glycine and taurine respectively.

14-1-5. *Steroid hormones.* Many animal hormones possess steroidal structures. These include the hormones of the adrenal cortex (cortisone and its derivatives) and the regulatory factors of sex and reproduction. A detailed discussion of these important substances and their functions is outside the scope of this book. Some aspects of relevance in food and biotechnology will be mentioned.

Interference with the natural course of sexual development of domestic animals has always been an accepted practice in agricultural husbandry. The purpose of such interference is often to accelerate growth, improve feed efficiency or modify the fat deposition pattern. Lately, massive administration of steroid hormones and physiologically related substances has been practised in poultry production instead of the more tedious and less efficient chirurgical caponizing. The most widely practised method of administration was the implantation of a capsule of the active substance in the head of the animal. The substance most widely used was *diethylstilbestrol*, which is not a natural hormone, but a synthetic substance with physiological properties similar to those of natural estrogens (female sex hormones). Although this particular application of diethylstilbestrol has been discontinued for public health reasons, the estrogen is still being used as a feed efficiency improver in beef cattle production.

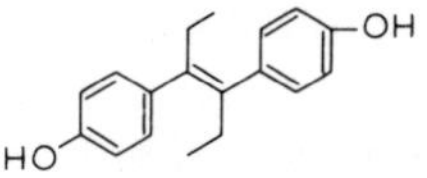

diethylstilbestrol: a synthelic estrogen

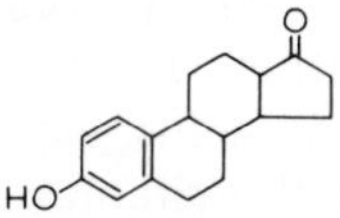

estrone: a natural estrogen

The hormones of the adrenal cortex are of primordial importance in the regulation of life processes and metabolism. Over 40 steroids have been identified in adrenal extracts. Only seven are hormones. They all possess ketonic carbonyl at C_3 and C_{20}. Their skeleton structure contains 21 carbons and the terminal C_{21} position is hydroxyl substituted. The structure of two active adrenal steroids, *cortisone* and *corticosterone* is shown below.

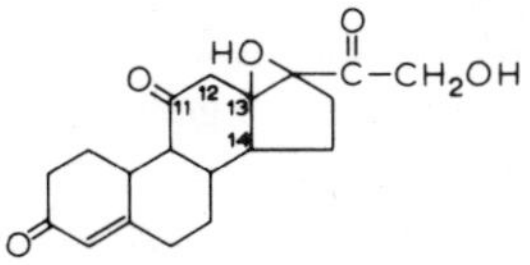

cortisone

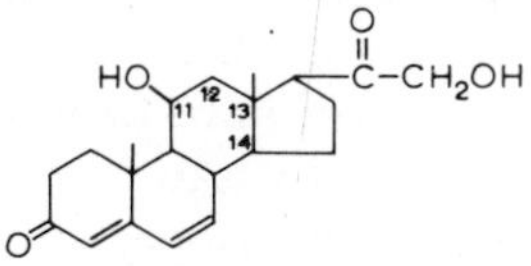

corticosterone

Discovery of its many-sided therapeutic properties brought cortisone into prominence as a "wonder drug" in the 1950s.

At first, cortisone isolated from glandular material or synthesized from bile acids was extremely expensive and of short supply. Chemical synthesis from bile acids required thirty-six steps. The most difficult step was the introduction of oxygen at the C_{-11} position. In 1952, researchers from two leading pharmaceutical companies in the U.S.A. reported on the ability of certain fungi to carry out this specific reaction in relatively high yields. This ability has been utilized since then in the synthesis of cortisone drugs from more easily available steroids. Besides its direct impact on the steroid industry, this finding has considerable historical importance, since it demonstrates the feasibility of using biological processes in order to induce highly specific molecular modifications.

14-2. TOCOPHEROLS, VITAMIN E

14-2-1. *Occurrence, structure.* Tocopherols are widely distributed fat soluble substances recognized as essential dietary factors. Their structure features an isoprenoid part attached to a polysubstituted phenol ring.

Testing the ability of chemically well defined diets to sustain normal life in animals is one of the important techniques of experimental nutrition. In the 1920s the list of known essential materials included proteins, carbohydrates, fats, vitamin A, many of the components of the vitamin B complex, ascorbic acid, vitamin D and a number of minerals. When diets containing all these in above-optimal quantities were fed to rats, certain abnormalities were observed. The females failed to give birth and the males developed atrophy of the testes. These observations led to the assumptions that an unknown vitamin, absent in these synthetic diets and essential for normal reproduction must exist. The unknown factor was named *Vitamin E.*

In 1936, vitamin E was isolated in a pure state and found to be the closely related substances α- and β-*tocopherol* (from the Greek *toko* = birth and *phero* = to bear, —ol indicating an alcohol). Since then, five more derivatives have been isolated from wheat germ oil, corn oil, soybean oil, etc. All tocopherols are derivatives of a 6-hydroxychroman with an isoprenoid sidechain in position 2, and they differ only in their substitution groups at carbon atoms 5, 7 and 8. The following are the structural formulae of some tocopherols:

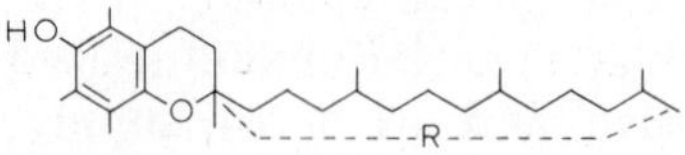

α-tocopherol (5,7,8-trimethyltocol)

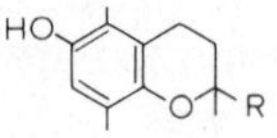

β-tocopherol (5,8-dimethyltocol)

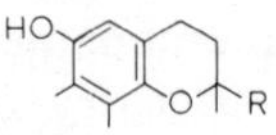

γ-tocopherol (7,8-dimethyltocol)

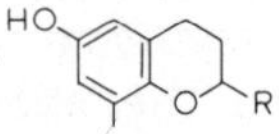

δ-tocopherol (8-monomethyltocol)

All tocopherols are oily liquids. The International Unit is 1 mg of tocopherol–acetate. As we shall see later, the adequate level of daily intake depends on other factors in the diet. However, in practical nutrition, vitamin E deficiency is extremely rare. Animal foods are generally poor sources of vitamin E. The best sources are wheat germ seed and leaf vegetables. Typical contents, in milligrams of total tocopherol per 100 grams of food, are: wheat germ 20, peas 17, wheat 3, corn 6, cabbage 5, spinach 3, eggs 2, beef 0.6. Edible vegetable oils contain high levels of tocopherols, in the range of 100 mg per 100 grams.

14-2-2. *Vitamin E function.* As we have seen before, vitamin E deficiency induces *sterility* in rats. While the effect of such deficiency in a female rat can be reversed by addition of vitamin E to the diet, the sterility of the male rat is irreversible. Although popular belief usually associates vitamin E with sterility alone, the biological function of tocopherols covers a number of other areas. The main effect of vitamin E deficiency in guinea pigs, rabbits, calves and a number of domestic animals is the *dystrophy of skeletal muscles.* In chicks severe vitamin E deficiency affects the brain and causes an irreversible pathological condition known as *encephalomalacia* or the "crazy chick disease".

The principal mechanism of vitamin E action in the body is believed to be associated with the antioxidant properties of tocopherols. Tocopherols are known to stop autoxidation of unsaturated lipids (see Chapter 15). Part of the vitamin E action may be restored by the addition of synthetic antioxidants and selenium to diets deficient in tocopherols. Furthermore, deficiency symptoms may be induced by increasing the level of unsaturated fatty acids in the diet. It is now known that the vitamin E requirement of an animal depends on the unsaturated fatty acid content of the food.

The possibility of the participation of tocopherols in some important enzymic process as a co-enzyme has been postulated as an additional mechanism of vitamin E action.

14-2-3. *Stability of tocopherols.* Tocopherols are easily oxidized. The mechanism of oxidation of α-tocopherol has been shown to follow the following scheme:

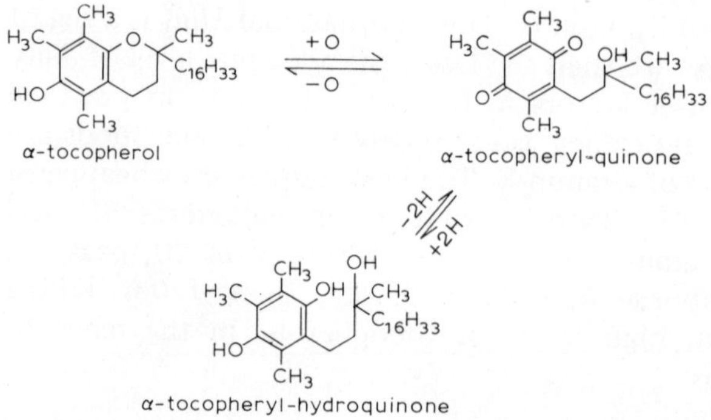

H_3C, CH_3, O, CH_3, $C_{16}H_{33}$, HO, CH_3

α-tocopherol

+O / −O

CH_3, H_3C, O, OH, CH_3, O, $C_{16}H_{33}$, CH_3

α-tocopheryl-quinone

−2H / +2H

CH_3, OH, H_3C, OH, CH_3, $C_{16}H_{33}$, HO, CH_3

α-tocopheryl-hydroquinone

Apart from their susceptibility to oxidation, tocopherols are very stable and can withstand high temperatures as well as treatment with acid and alkali. This is the reason for the excellent retention of tocopherol content in edible oils in process of their raffination (see Chapter 11). A certain proportion of the tocopherols is removed in the process of adsorbtion bleaching.

SELECTED BIBLIOGRAPHY TO CHAPTER 14

Dyke, S.F., The Chemistry of the Vitamins, Interscience Publishing, New York, 1965.

Florkin, M. and Stotz, E.H., Comprehensive Biochemistry, Vol. 10, Elsevier Publishing Co., Amsterdam, 1963.

Heftmann, E., Steroids, in Plant Biochemistry, Y. Bonner and J.E. Varner (Eds.), pp. 694-716, Academic Press, New York, 1965.

Heftmann, E., Steroid Biochemistry, Academic Press, New York, 1970.

McKerns, K.W., Steroid Hormones and Metabolism, Appleton-Century-Crofts, New York, 1969.

Morton, R.A., Fat Soluble Vitamins, Pergamon Press, Oxford, 1970.

CHAPTER 15

LIPID OXIDATION

15-1. INTRODUCTION

The occurrence of off-flavors, generally described as *"rancidity"*, in fat-containing foods is a common observation. The principal source of "rancidity" in foods is the autoxidation of lipid components. *Autoxidation* is defined as the spontaneous oxidation of a substance in contact with molecular oxygen.

Although the onset of "rancidity" is the most significant consequence of lipid autoxidation, flavor deterioration is not the only damage suffered by foods in this process. Color is also affected through accelerated browning reactions. The nutritional value is impaired and even toxicity may be induced. The texture may also change as a result of side reactions between proteins and the products of fat oxidation. In short, oxidative deterioration of lipids may be considered as a spoilage factor affecting all the aspects of food acceptability.

The lipid components most susceptible to autoxidation are the unsaturated fatty acids, especially those with more than one double bond. Although formally the process consists of a reaction between two molecular species (the lipid and oxygen), the number of possible pathways increases enormously in the course of the reaction. Even in the simplest model system consisting of one fatty acid and oxygen, a very large number of intermediates and final products are found and the system soon becomes very complex. In this respect, autoxidative deterioration of lipids resembles somewhat non-enzymic browning (Chapter 10).

15-2. MECHANISM

The *"hydroperoxide" course* proposed more than thirty years ago by Farmer and his co-workers, is accepted as the central mechanism of lipid autoxidation.

15-2-1. *General course.* According to Farmer's theory the reaction proceeds through a free radical mechanism consisting of the following steps (in our case, RH represents a molecule of lipid):

Step 1: Initiation

$$RH \xrightarrow{\text{activation}} R^{\circ} + H^{\circ} \quad (1)$$

Step 2: Propagation

$$R^{\circ} + O_2 \longrightarrow ROO^{\circ} \quad (2)$$

$$ROO + RH \longrightarrow R^{\circ} + ROOH \quad (3)$$

Step 3: Decomposition

$$ROOH \longrightarrow RO^{\circ} + OH^{\circ} \text{ (also } R^{\circ}, ROO^{\circ} \text{ etc.)} \quad (4)$$

Step 4: Termination

$$ROO^{\circ} + X \longrightarrow \text{Stable compounds} \quad (5)$$

In the first stage a few molecules of the lipid RH are sufficiently activated by heat, light or metal catalyst, etc., to decompose into the unstable free radicals $R^{\bullet}$ and $H^{\bullet}$. This free radical generation is by no means limited to lipids but can occur in any organic substance. Normally, the free radicals would quickly disappear by recombination into RH, RR, H_2, H_2O, etc., but in the presence of molecular oxygen the possibilities include an encounter between the free radical $R^{\bullet}$ and O_2, resulting in the peroxide radical $ROO^{\bullet}$. This radical then reacts with a fresh molecule of lipid, RH, producing the hydroperoxide ROOH and a free radical $R^{\bullet}$ through which the chain reaction is propagated. Now free radicals continue to be formed, without the help of the initial activator. The reaction proceeds and more lipid molecules are transformed to hydroperoxides. The reaction is terminated when free radicals combine with other free radicals or with free radical inactivators (represented by X), to yield stable compounds which accumulate in the system. The hydroperoxides enter a series of reactions leading to more free readicals and stable final products. These final products include short chain carbonylic compounds responsible for the rancid flavor and for side reactions leading to overall deterioration.

15-2-2. *Hydroperoxides.* The various regions of the lipid are not equally susceptible to activation. The methylenic group, adjacent to a double bond of a fatty acid, is particularly labile. In the light of this assumption Farmer concluded that oleic acid, for instance, would generate a free radical which can be represented as a hybrid of four resonant structures as shown below:

$$\overset{11}{R_1-CH_2}-\overset{10}{CH}=\overset{9}{CH}-\overset{8}{CH_2}-R_2 \quad (6)$$

$$\downarrow$$

$$\left.\begin{array}{ll} \text{I} & R_1-CH_2-CH=CH-\overset{\circ}{C}H-R_2 \\ \text{II} & R_1-\overset{\circ}{C}H-CH=CH-CH_2-R_2 \\ \text{III} & R_1-CH=CH-\overset{\circ}{C}H-CH_2-R_2 \\ \text{IV} & R_1-CH_2-\overset{\circ}{C}H-CH=CH-R_2 \end{array}\right\} + H^{\circ}$$

Accordingly, four hydroperoxides would be formed, having the —OOH group attached to positions 8, 9, 10 and 11. The theory has been confirmed by the detection of all four hydroperoxides in autoxidized oleate.

Similarly, linoleic acid yields three isomeric hydroperoxides (9, 11 and 13) and linolenic six (9, 11, 12, 13, 14, 16). Linoleic acid autoxidizes much more readily than does oleic acid, because in the former both double bonds activate the methylenic group located between them.

The hydroperoxides are important as the primary products of lipid autoxidation. They are in themselves non-volatile, odorless and tasteless. Their formation and accumulation, measured as the increase in the "peroxide value", indicates the progress of autoxidation, but not necessarily the appearance of rancidity.

15-2-3. *Degradation of hydroperoxides.* Hydroperoxides are relatively unstable. As their concentration in the system increases, they begin to decompose. One of the possible reactions is the monomolecular decomposition of hydroperoxides into an *alkoxy* and an *hydroxy* radical:

$$R_1-\underset{\displaystyle OOH}{\underset{|}{CH}}-R_2 \longrightarrow R_1-\underset{\displaystyle O^{\circ}}{\underset{|}{CH}}-R_2 + OH^{\circ} \quad (7)$$

Various possibilities exist for further reactions of the alkoxy radical:

(a) *Aldehyde generation*

$$R_1-\underset{\displaystyle O^{\circ}}{\underset{|}{CH}}-R_2 \longrightarrow R_1-\underset{\displaystyle O}{\underset{\|}{CH}} + R_2^{\circ} \quad (8)$$

Following this path, a short chain aldehyde is formed. For instance, the oleate peroxides would produce C_8, C_9, C_{10} and C_{11} aldehydes.

Attention is drawn to the equivalent formation of the new free radical $R_2^{\bullet}$, which would start its own chain reaction. The aldehyde itself may be oxidized to an acid, reduced to an alcohol, react with amine groups, etc. In the case of glycerides, the free radical R_2 is still attached to glycerol.

(b) *Reduction to an alcohol*

$$R_1-\underset{O^{\circ}}{\underset{|}{C}H}-R_2 + R_3H \longrightarrow R_1-\underset{OH}{\underset{|}{C}H}-R_2 + R_3^{\circ} \quad (9)$$

Here the alkoxy radical reacts with another lipid molecule, generating an alcohol and a free radical $R_3^{\bullet}$, which again participates in the propagation of the chain.

(c) *Formation of ketones.* The alkoxy radical could be oxidized by another free radical. This is a termination reaction.

$$R_1-\underset{O^{\circ}}{\underset{|}{C}H}-R_2 + R_3^{\circ} \longrightarrow R_1-\underset{O}{\underset{\|}{C}}-R_2 + R_3H \quad (10)$$

The monomolecular decomposition of hydroperoxides to alkoxy and hydroxy radicals seems to be the predominant route as long as the extent of oxidation is relatively low. At more advanced stages of the oxidation, bimolecular processes take over, as follows:

$$ROO\boxed{H + HO}OR \longrightarrow H_2O + ROO^{\circ} + RO^{\circ} \quad (11)$$

15-2-4. *Polymerization.* Formation of viscous, gum-like or even solid polymers (resins) is one of the consequences of lipid autoxidation. The "*drying*" of highly unsaturated oils used in paints is the result of such polymerization. Polymerization may occur through direct recombination of free radicals or through other reactions such as:

$$R_1-CH{=}CH-R_2 + R_3-OOH \longrightarrow R_1-\underset{OR_3}{\underset{|}{C}H}-\underset{OH}{\underset{|}{C}H}-R_2$$

15-3. KINETIC ASPECTS

The course of autoxidation in lipids is experimentally followed by measuring the accumulation of peroxides, the rate of oxygen uptake, the concentration of secondary reaction products or by organoleptic evaluation. Each one of these parameters may be indicative of a different aspect of oxidative deterioration.

"*Peroxide value*" is one of the most widely used concepts in lipid chemistry. It is a measure of the peroxide concentration of an oil, measured iodometrically or instrumentally, and expressed as "milliequivalents of peroxide oxygen per 100 grams of fat".

"*Rate of oxygen uptake*" is a more meaningful measure of the rate of oxidation. It is determined by somewhat complicated techniques similar to those used for the study of respiration rate in vitro.

Determination of decomposition products is meant principally to supply information on the build-up of flavorful compounds responsible for rancidity. These compounds may be identified by various chemical methods or physical techniques such as gas chromatography. However, one of the most widely used methods involves the reaction of the oxidized fat with *thiobarbituric acid* (TBA). TBA reacts with oxidized fats to give a red colored complex which can be measured spectrophotometrically. Although the pigment is the product of condensation of TBA with one specific substance, *malonic dialdehyde*, the TBA test usually correlates well with the degree of rancidity.

The organoleptic test detects the decomposition products in virtue of their odor and taste. Because of the very low flavor threshold values of many of these products (less than 1 ppm), the test is quite sensitive and useful.

When the peroxide value PV (or oxygen uptake) of an autoxidizing lipid is followed, a curve such as the one shown in Fig. 25 is observed. At first PV increases at a slow, uniform rate. As soon as the PV reaches a critical value, depending on the system, a sudden and drastic increase in rate is recorded. The first slow phase is termed "*induction period*". The autocatalytic nature of this course, i.e. the increase of reaction rate as the reaction proceeds, is explained on the basis of the free radical chain mechanism explained above. Rancidity usually begins to develop soon after the transition between the two phases. Obviously, at this stage oxygen adsorption rate is a better measure than PV, since at this point considerable peroxide decomposition occurs simultaneously with new peroxide formation.

The behavior depicted by Fig. 25 may be explained, in a somewhat oversimplified manner, as follows: During the induction period initiation and propagation (reactions 1, 2 and 3) occur. Since for each free radical which is transformed to a hydroperoxide one new free radical is formed, the reaction proceeds at a slow, uniform rate. As the concentration of hydroperoxides increases, hydroperoxide decomposition reactions take place at an increasing rate. These reactions generate more free radicals than needed for the

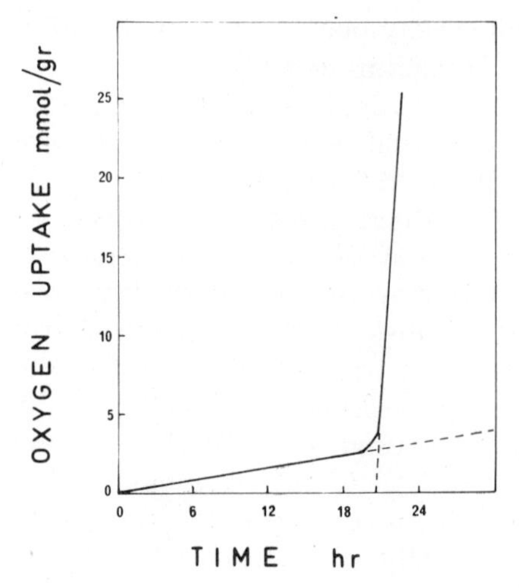

Fig. 25. Typical rate curve of lipid oxidation.

propagation of the chain reaction at a constant rate. Consequently, the reaction becomes autocatalytic. It is appropriate to note in this connection that many accelerating factors increase the rate of autoxidation, not by interfering with the initiation and propagation steps, but rather by enhancing the rate of peroxide decomposition. The appearance of perceptible rancidity at the transition point, or near it, is also explained by the decomposition of hydroperoxides to aldehydes and ketones which bear the characteristic rancid odor and taste.

The fundamental study of lipid oxidation kinetics, in terms of the rate constants relative to the successive step reactions, has received considerable attention. The interested reader is referred to a recent review by Labuza.

15-4. EFFECT OF ENVIRONMENTAL FACTORS

15-4-1. *Temperature.* The rate of autoxidation increases with temperature. The energy of activation is strongly dependent on temperature. At lower temperatures Q_{10} values larger than 2 are observed. At high temperature the Q_{10} value approaches 2, as in most chemical reactions. However, since high temperatures accelerate both the generation of free radicals and their disappearance, the rate–

temperature relationship may be expected to pass through a maximum, especially at high oxidation levels and high temperatures.

Temperature may affect not only the rate of autoxidation but the reaction mechanism as well. At lower temperatures the hydroperoxide course described above is the predominant mechanism. In this process there is no loss of unsaturation. At higher temperatures a considerable proportion of double bonds may undergo saturation.

15-4-2. *Light.* Fatty acids and their peroxides are colorless substances that do not absorb visible light. Thus, unless an accessory sensitizer is present, the effect of visible light on lipid autoxidation may be assumed to be unimportant. However, ultraviolet light is strongly absorbed by unsaturated compounds, especially if the double bonds are conjugated. Ultraviolet light may be a factor in the initiation of the chain reaction but its principal effect is attributed to the acceleration of peroxide decomposition. In daily practice, it is observed that "oxidized flavor" is induced more readily in bottled milk than in milk packed in opaque cartons. For the same reason, salad oil is often bottled in colored glass.

15-4-3. *Oxygen.* As long as oxygen is present in limited quantity, the rate of autoxidation increases with increasing oxygen pressure, until a constant oxidation rate is reached beyond a given pressure. At low oxygen pressures the reaction rate has been found to be proportional to oxygen pressure. More recent work, however, has discovered linear relationship between the reciprocal of oxidation rate and the reciprocal of oxygen tension. According to Karel, the specific surface of the lipid system is more important than oxygen tension. In systems with very large surface/volume ratio, such as dehydrated foods or a model system consisting of a lipid adsorbed on powdered cellulose, the rate of oxidation is very rapid and almost independent of oxygen pressure. Exclusion of air (vacuum or nitrogen packs, use of packaging materials with low oxygen permeability) is among the practical methods commonly used for the retardation of oxidative deterioration of fatty foods (nuts, coffee, baked goods). The effectiveness of this measure may be limited, however, in the case of very dispersed lipid systems such as instant soup powders, milled feedstuffs, etc.

15-4-4. *Moisture.* The effect of water activity on the rate of lipid oxidation is complex. Rancidity develops more rapidly both at very high and very low moisture levels. Maximum stability is observed at intermediate moisture levels corresponding to monolayer values. The protective effect of water at monolayer values has been explained in several ways:

(a) Interaction with metal catalysts.
(b) Retardation of oxygen transport to the lipid phase.
(c) Stabilization of peroxides by hydrogen bonding.

It is interesting to note that rancidity is particularly important in frozen foods (fish, meat), because separation of water as ice crystals apparently deprives the lipids from the protective effect of the moisture films.

15-4-5. *Ionizing radiations.* High energy irradiation (β and γ radiation) has been extensively studied as a food preservation method. During the 1950s and early 60s, this was probably the most frequently treated subject of the food science literature. One of the striking effects of high energy irradiation of foods is a marked increase of susceptibility to oxidative rancidity. This is usually explained on the basis of radiation-induced free radical generation. Highly saturated fats, which would normally exhibit higher stability, are strongly affected by irradiation. Thus, rancidity can be a problem in irradiation-preserved lard and bacon. That the effect of radiation is mainly connected with increased free radical formation is evident from the observation that rancidity develops not during irradiation but upon storage of the irradiated food.

15-4-6. *Catalysts.* Ions of heavy metals are powerful catalysts of lipid oxidation. They shorten the induction period and increase the reaction rate. Most effective are metals that can exist in two or more states of oxidation and can easily pass from one state to another, e.g. iron, copper, manganese, etc. Most foods and even refined fats and oils contain these metals at concentration levels above those needed for effective catalysis, which are in the order of ppm values or less.

The main effect of these trace metals is to increase the rate of hydroperoxide decomposition, and hence the rate of free radical generation. This action may be represented as follows:

$$ROOH + M^{n+} \longrightarrow RO^{\circ} + OH^{-} + M^{(n+1)+} \qquad (12)$$

$$M^{(n+1)+} + ROOH \longrightarrow ROO^{\circ} + H^{+} + M^{n+} \qquad (13)$$

Thus, hydroperoxides are decomposed and free radicals $RO^{\bullet}$ and $ROO^{\bullet}$ are formed as the metal oscillates between its two oxidation states. However, for most metals, reaction 12 is much faster than reaction 13.

Metal ions may also participate in the initiation process by activating an unoxidized molecule of lipid or a molecule of oxygen. They can also catalyze termination processes and act as inhibitors. However, in comparison to the hydroperoxide decomposition action, these other effects are rarely significant.

The source of heavy metal ions in foods may be contamination (equipment, piping, packaging materials, environmental contaminants) or natural food components. As far as lipid oxidation catalysis is concerned, the most important metal-containing natural food components are the metallo-porphyrin substances (hematin compounds), such as hemoglobin, myoglobin and cytochromes (see Chapter 18). Catalysis by hematin is especially important in frozen meat and fish.

The exact mechanism of hematin catalysis is not known. The action of hematins as lipid oxidation catalysts might be due to the oxidation-reduction of the metal atom in the heme molecule. According to Tappel, however, there is no change of valence of the metal during hematin-catalyzed lipid oxidation. He suggests a mechanism which involves formation of a hematin-peroxide complex. Decomposition of the complex yields free radicals which propagate the chain. The mechanism may be pictured as follows:

$$\text{(N}_4\text{)Fe–OH} + ROOH \longrightarrow \text{(N}_4\text{)Fe–OOR} + H_2O$$

$$\text{(N}_4\text{)Fe–OOR} \longrightarrow \text{(N}_4\text{)Fe–O}^{\circ} + RO^{\circ}$$

$$\text{(N}_4\text{)Fe–O}^{\circ} + RH \longrightarrow \text{(N}_4\text{)Fe–OH} + R^{\circ}$$

Hematin-catalyzed lipid oxidation in vivo may cause pathogenic conditions. In the normally functioning organism this is counteracted by vitamin E (see Chapter 14).

Another important catalyst of lipid oxidation in some foods is the enzyme *lipoxidase.* Lipoxidase catalyzes specifically the direct oxidation of poly-unsaturated fatty acids containing a cis-cis 1,4-*pentadiene* group (e.g. linoleic and linolenic acid). The enzyme is found in oilseeds, legumes, cereals and leaves. If it is not inactivated by heat (blanching), lipoxidase may cause rapid development of off-flavors in frozen and dehydrated vegetables, especially in peas. The oxidative destruction of carotenoids in dehydrated vegetables and alfalfa (see Chapter 12) is usually due to lipoxidase catalyzed coupled oxidation. Bleaching of wheat flour is also due to the same phenomenon and may be accelerated by boosting the natural lipoxidase activity of wheat by addition of untoasted soy flour or other lipoxidase-rich preparations.

15-4-7. *Antioxidants.* Antioxidants are substances that retard autoxidation. In theory, a substance may act as an antioxidant in a variety of ways, e.g. competitive binding of oxygen, retardation of the initiation step, blocking of propagation by destroying or binding free radicals, inhibition of catalysts, stabilization of hydroperoxides, etc. All these and other mechanisms are found in food systems but the most important one seems to be blockage of propagation. In this process, the antioxidant AH acts as a hydrogen donor to a free radical such as $ROO^{\bullet}$ or $R^{\bullet}$:

$$AH + ROO^{\circ} \longrightarrow ROOH + A^{\circ} \quad (14)$$

$$\text{or} \quad AH + R^{\circ} \longrightarrow RH + A^{\circ} \quad (15)$$

The antioxidant free radical $A^{\bullet}$ is inactive, i.e. it does not start a chain propagation process, but rather enters some termination reaction such as:

$$A^{\circ} + A^{\circ} \longrightarrow A - A \quad (16)$$

$$\text{or} \quad A^{\circ} + ROO^{\circ} \longrightarrow ROOA$$

The antioxidant may also be regenerated if a secondary hydrogen donor BH is present, and if the oxidation–reduction potential of the following reaction is favorable:

$$A^{\circ} + BH \longrightarrow AH + B^{\circ} \quad (17)$$

Antioxidant action increases the length of the induction period, as shown in Fig. 26. The induction period increment is roughly proportional to the concentration of antioxidant up to a certain level. Excess concentration of antioxidant is ineffective or may even cause reversion of the protective effect.

One of the principal classes of antioxidant are the natural or synthetic phenolic compounds. Several of the natural phenols described in Chapter 16 (catechol derivatives, tannins, gossypol,

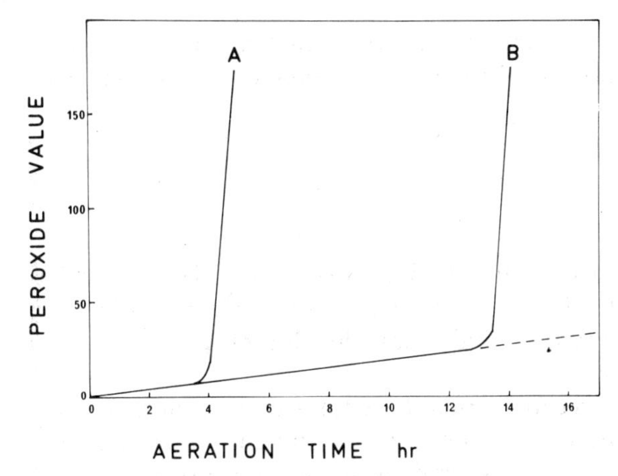

Fig. 26. Effect of antioxidant on lipid oxidation rate: (a) fat without antioxidant; (b) fat with antioxidant.

gallic acid) and the tocopherols discussed in Chapter 14 are potent antioxidants. Synthetic phenolic antioxidants approved for food use include *butylated hydroxyanisol* (BHA), *butylated hydroxytoluene* (BHT) and *propyl gallate* (PG).

BHA BHT PG

BHT and BHA are quite volatile at frying temperatures. PG forms dark compounds with iron ions.

BHA, BHT and PG are recognized as safe food additives in most countries. They are used at concentrations of up to 200 ppm on the basis of fat. Other synthetic phenolic antioxidants include *nordihydroguaretic acid* (NDGA), which is now prohibited in some countries, and *ethoxyquin* which is mainly used in animal feeds for the protection of carotenoids.

NDGA ethoxyquin

The stability of the phenolic antioxidant free radical (A in reaction 14) is explained on the basis of resonance of the phenoxy system, as follows:

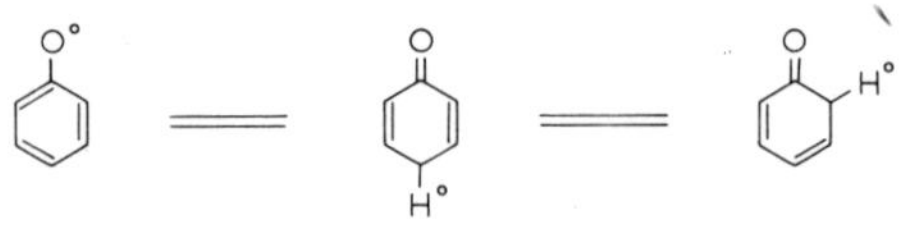

The tertiary butyl groups attached to BHA and BHT somewhat enhance the stability of the corresponding phenoxy radical by introducing steric hindrance which prevents interaction with lipid molecules RH (propagation).

The empirical use of natural phenolic compounds as antioxidants is very old. The popularity of smoking and spicing in the home — preservation of meat, fish, cheese and other fat-rich foods — may be due, at least partly, to the recognition of the rancidity retarding effect of these treatments.

Among the natural antioxidants, tocopherols deserve special attention. As we have seen in the previous chapter, the in vivo antioxidant activity of the tocopherols is the main, and perhaps only reason, for their vitamin E activity. In technology, the natural tocopherol content of many oils is important for their stability. However, on an equal weight basis, tocopherols are less effective than synthetic phenolic antioxidants, and much more expensive. For this reason they are not used as intentional antioxidant additives.

Some bioflavonoids (see Chapter 16) such as *quercetin* have been reported to possess lipid antioxidant activity; however, they are not used in practice.

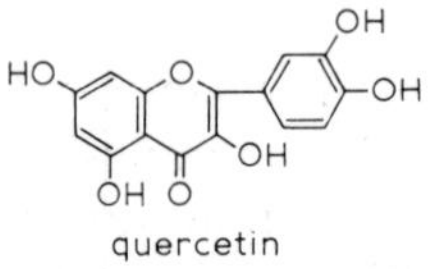

quercetin

The antioxidants discussed so far are "*primary antioxidants*". They interfere directly with the free radical propagation process and they block the chain reaction. There exist a number of other substances which have little direct effect on the autoxidation of lipids but are able to enhance considerably the action of primary antioxidants. These substances are termed "*synergists*". One of the best known and most widely used synergists is citric acid. Its indirect action is believed to be due to its ability to form stable complexes with pro-oxidant metal ions. In virtue of its poly-carboxylic and α-hydroxylic structure, citric acid is a potent metal chelating agent.

```
   CH2-COOH
    \
HO-C-COOH
    /
   CH2-COOH
 citric acid
```

Its direct activating effect on phenolic primary antioxidants is probably non-specific, and due to its acidic character.

Crude lecithin has been used as an antioxidant. A good part of its activity is probably due to the presence of tocopherols in the crude material. Purified lecithin is less active as a primary antioxidant but may act as a synergist, in virtue of the metal ion binding effect of its phosphoric acid group.

15-5. LIPID AUTOXIDATION IN FOOD SYSTEMS

15-5-1. *Oxidized flavors.* The immediately recognizable effect of lipid oxidation in foods is the development of undesirable odors and off-flavors. The chemical identity of a large number of "rancid" products of lipid oxidation has been determined. These are largely short-chain carbonylic compounds formed as a result of peroxide decomposition. The overall organoleptic nature of rancidity depends somewhat on the system. The rancidity of low-moisture foods is usually described as "old oil" or "tallow-like". On the other hand, lipid oxidation in water-rich foods such as milk results in "card-board-like" off-flavors, known as "oxidized milk flavor".

"*Flavor reversion*" is another oxidative deterioration process of great importance in some vegetable oils, such as soybean oil. Freshly refined soybean oil is practically tasteless. However, upon storage under improper conditions (extensive exposure to air, high temperature), off-flavors ranging from "beany" to "fish-like" are quickly formed. The term "reversion" implies that the refined oil reverts to its raw, unrefined form. This implication is incorrect, as the "reversed" flavor is due to newly formed compounds, unrelated to the flavor-bearing components of the raw oil.

Flavor reversion is usually due to the autoxidation of linoleic acid. It is therefore characteristic of oils with a relatively high poly-unsaturated acid content (linseed, soybean, rapeseed). The "reverted" flavors are due to unsaturated aldehydes, some of which are said to have flavor threshold values below one part in ten million. For this reason, flavor reversion is perceptible at the very early stages of oxidation at relatively low levels of peroxide value.

15-5-2. *Effect on color.* Lipid oxidations may affect indirectly the color of foods. In systems containing carotenoids, propagation of the lipid oxidation chain through free radicals may cause oxidative destruction of the carotenoid pigments. As mentioned before, this type of deterioration is important in dehydrated vegetables and usually involves the catalytic action of lipoxidases. Maillard type browning reactions may take place between protein and the carbonylic breakdown products of lipid oxidation. This type of reaction is believed to be the cause of yellowing (rusting) in frozen and pickled fish products.

15-5-3. *Effect on texture.* The interaction between proteins and the products of lipid oxidation may result in changes of texture. The mechanism of interaction involves propagation of the free radical chain to the protein system. Various groups in the protein molecule are capable of converting to free radicals by losing a hydrogen atom to a free radical of lipid origin. The protein "free radicals" thus formed tend to combine by cross-linkage. Toughness in frozen fish has been attributed to the aggregation of myosin molecules as a result of attack by oxidized lipid free radicals. Similarities between the effects on oxidized fats and those of ionizing irradiation on protein, support the free radical attack theory. The site most susceptible to such attack in the protein molecule seems to be the NH group of histidine.

15-5-4. *Lipid oxidation at high temperature.* The subject is of interest in connection with food processing operations involving high temperatures: toasting, roasting, baking, frying. The most important characteristics of heated oils are the following:

(a) Despite the accelerated rate of oxidation, peroxide values are usually very low, due to the rapid decomposition of the peroxides formed.

(b) Unlike fats oxidized at low temperatures, the flavor of heated oils is not rancid or otherwise objectionable. On the contrary, their taste and odor are well accepted. This may be due to the elimination of the volatile breakdown products by evaporation. In the case of frying, such elimination is even more effective due to the steam distillation effect.

(c) On the other hand, polymerization is one of the predominant termination processes. The viscosity of oils increases considerably in the process of heating.

(d) The degree of unsaturation, measured as the "iodine value" decreases sensibly, indicating direct saturation of the double bonds. Poly-unsaturated acids are affected first.

(e) Hydrolysis of the fat occurs and fatty acids are liberated, especially in the process of frying.

15-5-5. *Toxicity of oxidized fats.* Studies on the biological effects of oxidized lipids on their various components have been carried out for many years. Yet the results are not conclusive. Massive ingestion of highly oxidized fats or concentrated fractions containing peroxides, or their decomposition products, has been reported to cause disturbances ranging from growth inhibition to carcinogenesis. In most cases, however, the levels of intake necessary to cause such disturbances was unrealistically high, as compared to the expected level of voluntary intake of rancid foods. The situation may be different in the case of heated and polymerized oils, where objectionable taste does not exist as a built-in warning factor.

SELECTED BIBLIOGRAPHY TO CHAPTER 15

Farmer, E.H., Bloomfield, G.F., Sundralingham, A. and Stutton, D.A., Trans. Faraday Soc., **38** (1972), 348.

Johnson, F.C., A Critical Review of the Safety of Phenolic Antioxidants in Foods, Critical Reviews in Food Technology, **2**, No. 3 (1971), 267.

Labuza, T.P., Kinetics of Lipid Oxidation in Foods, Critical Reviews in Food Technology, **2**, No. 3 (1971), 355.

Lundberg, W.O. (Ed.), Autoxidation and Antioxidants, 2 Vols., Interscience Publishers, New York, 1961.

Marcuse, R. (Ed.), Metal Catalyzed Lipid Oxidation, Sik., Goteburg, 1968.

Mattil, Karl F., Norris, F.A., Stirton, A.J. and Swern, D., Bailey's Industrial Oil and Fat Products, Interscience Publishers, New York, 1964.

Tappel, A.L., Schultz, H.W., Day, E.A. and Sinnuber, R. (Eds.), Lipids and their Oxidation, Avi. Pub. Co., Westport, Conn., 1962.

Weiss, T.J., Food Oils and their Uses, Avi. Pub. Co., Westport, Conn., 1970.

CHAPTER 16

PHENOLIC COMPOUNDS

16-1. OCCURRENCE

Living matter contains a very large number of phenolic and polyphenolic compounds which, although of widespread occurrence and considerable importance, do not account for more than a minute fraction of the organic substance of plants and animals. Of special interest are the so-called "*plant phenolics*". One group of natural phenolic compounds, the *anthocyanins*, are water soluble pigments widely distributed in the plant kingdom and responsible for the beautiful color of many flowers, fruits and vegetables. Another group, the *anthoxanthins*, include bitter principles; the recognition of the biological functions of some compounds of this group has earned them the name of "*bioflavonoids*". The *catechins* are responsible for much of the taste in tea and wine. The *tannins*, plant polyphenols of relatively high molecular weight, are astringent components of many foods and their ability to cause coagulation of proteins is the basis for their use as "tanning agents" in the leather industry. Several phenolics of vegetal and fungal origin are important toxic substances. Phenolic substrates in vegetables and fruits are involved in the well known spoilage phenomenon, "*enzymic browning*". At the same time, the stability of many foods depends on the presence of certain phenolic compounds acting as natural *antioxidants* (see Chapter 15). The antioxidant action of phenolics in vivo is of utmost interest in biology.

16-2. SIMPLE PHENOLICS

By this term we designate phenolic substances containing a single benzene ring. The phenolic amino acids *tyrosine*, *phenylalanine* and *tryptophane* have been mentioned earlier (Section 2-3). These, together with their simple derivatives, occur widely in foods and living tissue.

The C_6—C_3 phenylpropane pattern of phenylalanine and tyrosine is found in a number of natural phenolics. *Cinnamic acid* is the main constituent of Styrax balsam used in medicine. Its aldehyde, cinnamyl aldehyde, is found in cinnamon. Ortho- and para-hydroxy cinnamic acids are known as *o*- and *p-coumaric acids.* The stable lactone of *o*-coumaric acid, coumarin and its sugar esters, are widely distributed in plants.

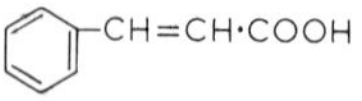

CH=CH·COOH

cinnamic acid

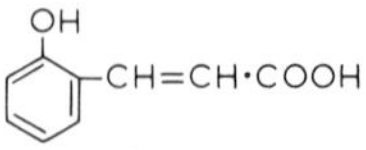

OH
CH=CH·COOH

o-coumaric acid

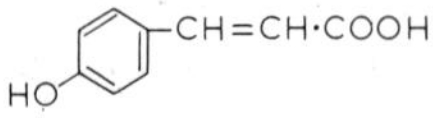

HO
CH=CH·COOH

p-coumaric acid

O
O

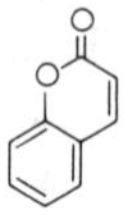

coumarin

Caffeic acid is found in coffee beans and digitalis. The ester of caffeic acid and *quinic acid* (acting as an alcohol) is *chlorogenic acid*, a substance of wide distribution in vegetables and fruits (potatoes, apples, pears). Its importance in enzymic browning will be discussed in the next chapter.

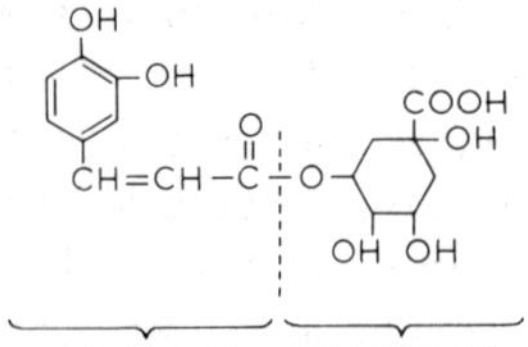

OH
OH
COOH
OH
CH=CH−C−O
OH OH

caffeic acid quinic acid

chlorogenic acid

The phenylpropane pattern is the principal structural unit of *lignin*, a phenolic macromolecular substance of wide occurrence as a major constituent of plant cell walls. Lignin accompanies cellulose and hemicellulose. It accounts for some 25% of most woods.

Some aromatic acids are found in the free state, such as benzoic acid in cranberries; some others in the form of esters, such as methyl salicylate in plums, concord grapes and acacia flowers.

Gallic acid is an important constituent of tannins. Its esters are widely distributed.

OH
HO OH
COOH

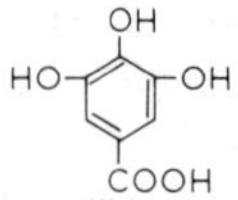

gallic acid

16-3. FLAVONOID COMPOUNDS

This large and important group encompasses the anthocyanins, anthoxanthins, leuco-anthocyanins and catechins. They all possess a basic structure consisting of a $C_6—C_3—C_6$ skeleton, and therefore are somewhat related to the phenylpropanoids discussed previously.

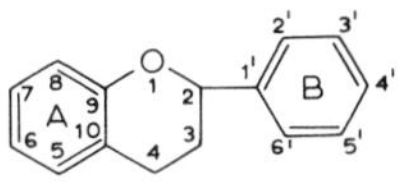

C_{15} skeleton of flavonoids

16-3-1. *Anthocyanins.* This is a very large group of plant phenolics, first studied by Richard Willstatter and later by Karrer and Sir Robert Robinson. These are the water soluble red, blue, purple pigments of flowers, fruit and vegetables. Their name comes from a term proposed by Marquart in 1835, to denote the blue substance of *Centaurea cyanus*, the cornflower. Anthocyanins are glycosides and when hydrolyzed, they yield a sugar and an aglycone, called *anthocyanidine.* The carbohydrate residues most frequently encountered are glucose, rhamnose, galactose and gentiobiose (see Chapters 6 and 7). The anthocyanidins have a common basic structure consisting of a benzopyrylium nucleus and a phenol ring, the two together being called *flavylium*:

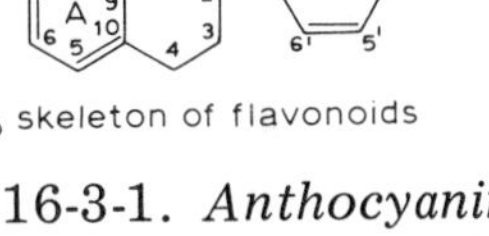

flavilium ion,
the parent structure of anthocyanidins

Because of the trivalent position of the oxygen, flavylium is a cation. The position at carbon atoms 3, 5 and 7 are hydroxyl substituted, the sugar moiety being usually attached to the hydroxyl at position 3. From this parent structural formula, anthocyanidins are derived by diverse substitution in the phenol ring B.

All the names of anthocyanidin are derived from the names of plants. *Pelargonidin* is found in the strawberry; *cyanidin* in the purple fig, almond, mulberry, sweet cherry and elderberry; *delphinidin* in pomegranate and eggplant.

The purple pigment of beetroot, *betanin*, was long thought to be an "unusual anthocyanin" as it contains nitrogen. The structure of betanin is now known to be quite different from that of flavonoid pigments.

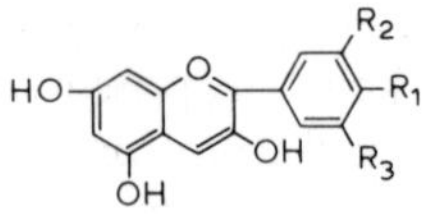

pelargonidin $R_1 = OH$, $R_2 = R_3 = H$
cyanidin $R_1 = R_2 = OH$, $R_3 = H$
delphinidin $R_1 = R_2 = R_3 = OH$

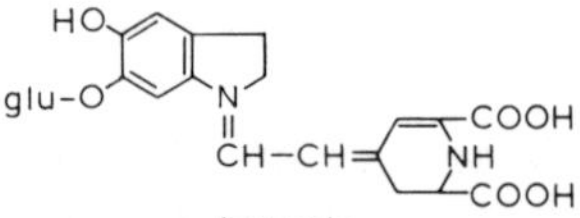

betanin

The anthocyanin pigments change their color with a change in pH. *Cyanine* (the 3,5-diglucoside of cyanidin) for example, is red in acidic solution, purple at neutral pH and blue in alkaline medium. The change in color is assumed to be associated with a change in configuration as shown below:

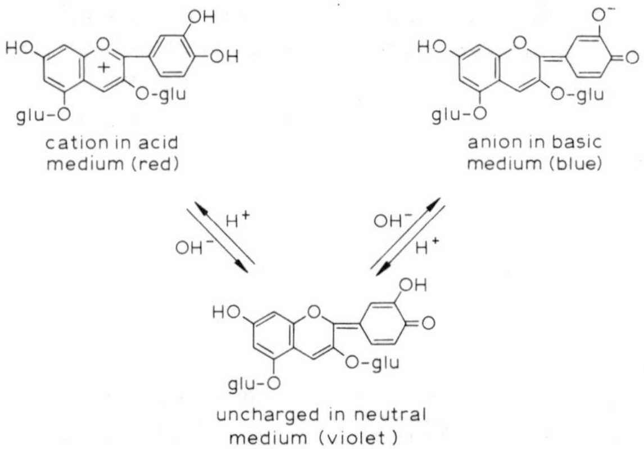

cation in acid medium (red)

anion in basic medium (blue)

uncharged in neutral medium (violet)

The pH dependence of the color and color stability of anthocyanin pigments is of some importance in food processing. Previously, the pH of the plant sap was believed to determine the color of flowers. Other mechanisms, including the formation of metal complexes, are now thought to be responsible for flower color variation.

Anthocyanin pigments are of utmost importance in red wines. It is interesting to note that in most varieties of "red" wine grapes, the pigment is located almost exclusively in the skin while the flesh of the berries is colorless. If the grapes are crushed and the "must" is filtered immediately, the resulting juice is colorless and suitable for the manufacture of white wines. In the production of red wines the juice is fermented in contact with the skins. The pigments diffuse into the juice and equilibrium is reached between juice and skins. Diffusion of anthocyanin pigments may be accelerated by heating the must before fermentation.

The most serious problem which is posed by anthocyanin pigments in food technology is their lack of chemical stability. Hydrolytic cleavage of the glycoside does not destroy the color but the aglycon being usually less soluble than the glycoside, precipitation of the pigment can occur. This is observed in the ageing of red wines. Reduction of anthocyanin destroys their color. Such reductive bleaching occurs in red berries, strawberries, etc., packed in unlacquered (plain) tin cans. The reducing conditions prevailing during the alcoholic fermentation of grape must is one of the sources of color loss in red wines. Sulfur dioxide bleaches the anthocyanin pigments. The discoloration is not due to reduction or a change in pH as previously postulated, but rather involves an addition reaction between bisulfite and the anthocyanidin. If SO_2 is removed by evaporation or by a strong sulfite binding agent such as formaldehyde, the color is partially restored.

Fungi and plants apparently possess enzyme systems capable of bleaching anthocyanin pigments. The exact nature of the enzyme in these systems is not known. The reaction apparently involves hydrolysis of the glycosidic bond followed by some oxidative degradation of the aglycone.

16-3-2. *Leucoanthocyanins.* This group of plant phenolics was long overlooked until 1920 when Rosenheim isolated one of them from grapes. Their name is due to the fact that they are colorless but produce anthocyanidins when treated with boiling hydrochloric acid. However, leucoanthocyanins are not leuco-bases of anthocyanins. Nor are they colorless precursors in the biosynthesis of plant pigments.

The basis of leucoanthocyanidins is flavon 3,4-diol.

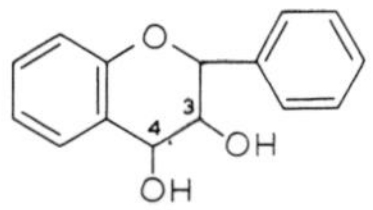

O
3
4
OH
OH

flavan - 3,4 - diol structure

Leucoanthocyanidins, their glycosides (leucoanthocyanins) and polymeric derivatives, are widely distributed in woody plants. Some are named after the anthocyanidin to which they give rise (*leucocyanidin, leucodelphinidin, leucopelargonidin*). Others are named according to the source from which they were first isolated, e.g. *melacacidin* from Acacia melanoxylon (Australian blackwood).

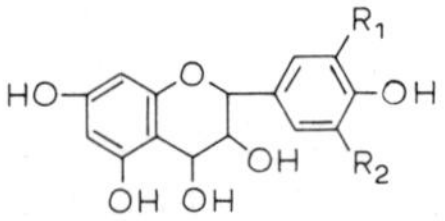

leucocyanidin	R_1 = OH, R_2 = H
leucodelphinidin	R_1 = OH, R_2 = OH
leucopelargonidin	R_1 = H, R_2 = H

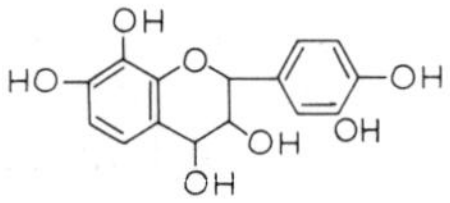

melacacidin

16-3-3. *Anthoxanthins.* The other group of glycosides, the *anthoxanthins*, differ from anthocyanins and leucoanthocyanins in the degree of oxidation of the aliphatic fragment in the $C_6—C_3—C_6$ skeleton. The aglycones of the anthoxanthins consist of a benzopyrone nucleus.

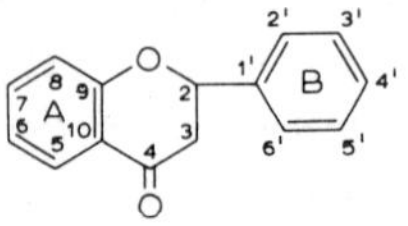

parent structure of anthoxanthius

The following are the various types of flavonoids (R indicates the position of the sugar residue). Apart from isoflavones and xanthones, the second aromatic ring (B) is always attached to the carbon atom in position 2, as in the anthocyanins.

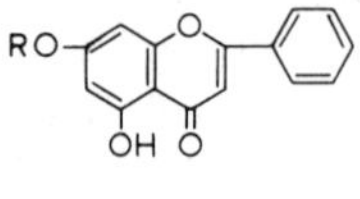

flavones

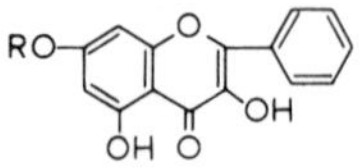

flavonols

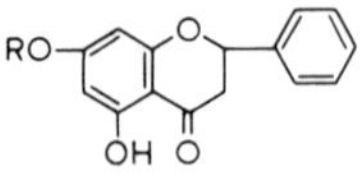

flavanones

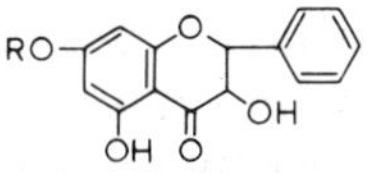

flavanonols

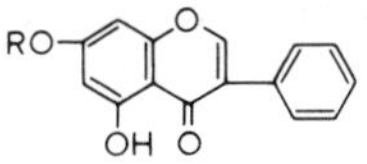

isoflavones

In the above formulae the sugar moiety is shown attached at position 7 of the A ring. This is indeed the most frequently encountered position of the glycosidic bond. However, 3, 5 and 4′-glycosides are also found often. 3-Glycosides of flavonols are particularly frequent. Unusual flavonone glycosides, known as the C-glycosides, have been found in citrus fruit. In these compounds the bond between aglycon and sugar is a direct C—C bond, without the normal oxygen bridge.

A few examples of anthoxanthins will make matters clearer. The flavonol *rutin*, found in tea leaves and also in other plants, consists of the aglycone *quercetin* and a combination of two sugars, a methyl pentose (rhamnose) and an hexose (glucose). The two sugars bound together are called *rutinose* and are linked to the aglycone in position 3, as is usual with most flavonols.

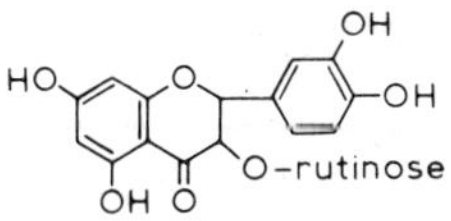

rutin

A quite similar aglycone, which is, however, a flavonone, is *eriodictyol*, the constituent of the glycoside *eriodictin*. Eriodictyol also has two hydroxyl groups in the B ring and can be converted into quercetin by amyl nitrite in the presence of hydrochloric and sulfuric acids, showing an example of converting, chemically, a flavonone into a flavonol.

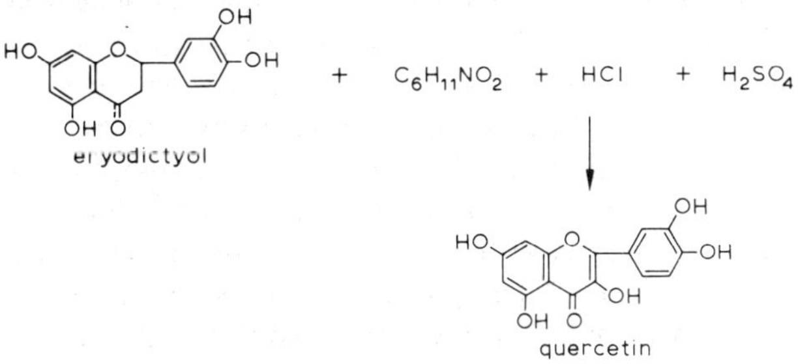

eryodictyol + $C_6H_{11}NO_2$ + HCl + H_2SO_4 → quercetin

A very widely distributed flavonone is *hesperidin* (from the Greek *Hesperides* — citrus fruits) also containing the disaccharide rutinose (rhamnose + glucose) at position 7. Hesperidin is only slightly soluble in water and somewhat more so in alcohol. M.p. = 252° C. Its aglycone, *hesperitin*, melts at 227° C.

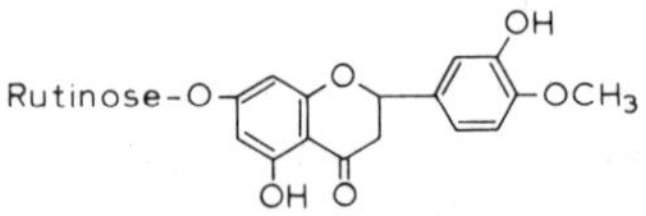

hesperidin

Hesperidin, and probably also all other flavonoids, is very soluble in alkali. At pH 11 to 12, the inner ring is opened forming *chalcones*:

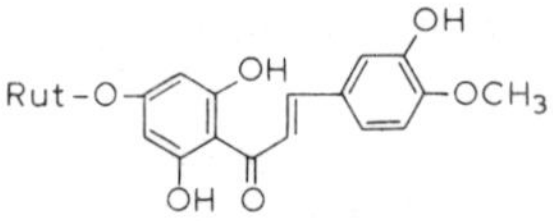

hesperidin chalcone

In contrast to anthocyanins, most of the flavonoids are colorless or only very slightly colored; however, their chalcones are yellow to strongly brown in color. There is ample reason to assume that in their chalcone form the flavonoids are salts of the corresponding alkali. At an acid pH the chalcones return to their former closed structure, thereby losing their color.

In order to stabilize and isolate the chalcones, it is customary to methylate the newly formed hydroxyl in position 1. Such methylated chalcones are extremely soluble in water.

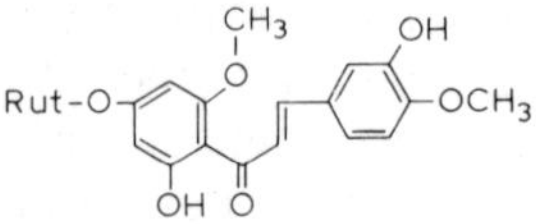

hesperidin methyl chalcone

Many anthoxanthin glycosides are bitter. However, cases where the bitterness of a food is due to flavonoids are not as frequent as previously assumed. In a number of cases, the occurrence of bitterness, formerly assigned to flavonoids, is now known to be due to other compounds, such as limonins (Chapter 13). One member of the flavonoid family, the glycoside *naringin*, found particularly in grapefruit and in sour oranges, is extremely bitter and imparts the specific bitterness to this fruit. Naringin (from the Sanskrit word "*naringi*" for orange), when pure, is a white crystalline glucoside, soluble in alcohol, acetone and only to the extent of 1 part in 2000 in water at 20° C. Its bitterness is even greater than that of quinine and it can be detected in dilutions of 1 : 50,000. When dried at 110° C, it melts at 171° C, but when crystallized from water with its additional 6 molecules of water it melts at 83° C. When dissolved in ethanol, naringin is levorotatory, $[\alpha]_D^{18} = -65°, 2'$. Its aglycone is called naringenin and the structural formula is as follows:

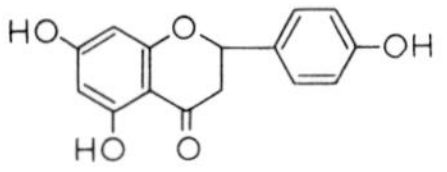

naringenin

The sugar moiety of naringin is neohesperidose, a rhamnosyl–glucose as rutinose. The rhamnose–glucose bond is $\alpha 1 \rightarrow 2$ in neohesperidose, but $\alpha 1 \rightarrow 6$ in rutinose.

A possible avenue of utilization of citrus flavonoids in foods is based on the recent development of "*dihydrochalcone sweeteners*". In their effort to elucidate the relationship between structure and taste of anthoxanthin flavonoids, Horowitz and Gentili discovered that when the intensely bitter naringin is converted to its chalcone and then reduced by hydrogen, an intensely sweet substance, *naringin dihydrochalcone* is formed. The dihydrochalcone of *neohesperidin* was found to be even sweeter.

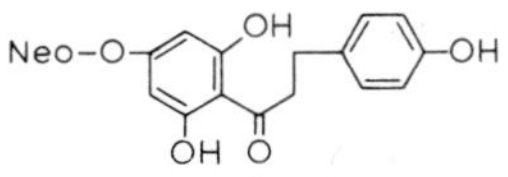

naringin dihydrochalcone

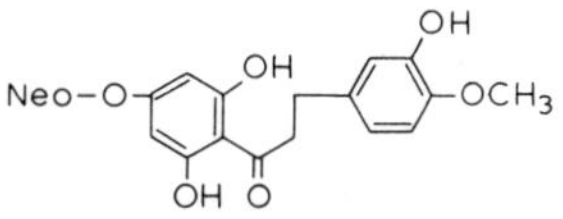

neohesperidin dihydrochalcone

Neohesperidin dihydrochalcone is said to be 2000 times sweeter than sucrose. Considerable research effort was devoted in the past ten years to the development of commercial sweeteners based on flavonoid dihydrochalcones, especially on neohesperidin DHC. Neohesperidin is a minor constituent of the flavonoid system in some varieties of citrus such as Seville oranges. However, it can be obtained from naringin. The conversion involves treatment of naringin with alkali under conditions that cause splitting of the B ring, followed by condensation with isovanillin.

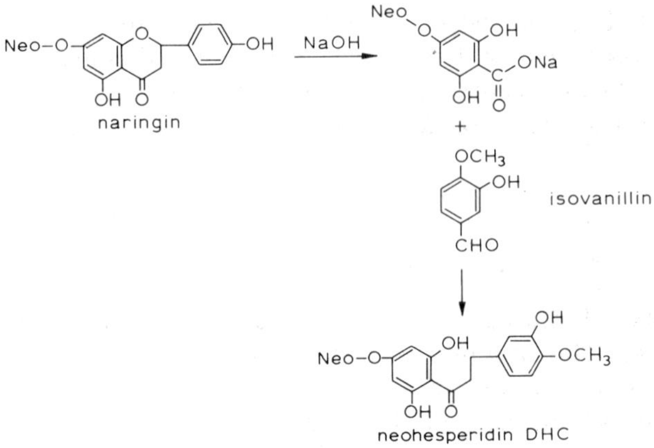

NaOH

naringin

isovanillin

neohesperidin DHC

At present, DHC sweeteners are awaiting official approval as safe food additives.

Before leaving the subject of anthoxanthin flavonoids, it is appropriate to mention the question of their antioxidant action. The relatively high ascorbic acid retention in processed citrus products led to the assumption that citrus fruits may contain effective antioxidant systems. Since the antioxidant action could be often traced back to the pulp and peel fraction, and in view of their phenolic nature, anthoxanthins were generally recognized as the active antioxidants. However, in the light of more recent work with purified model solutions, it seems that the antioxidant system protecting ascorbic acid and terpenes in citrus products does not involve the major citrus flavonoids naringin, hesperidin or their aglycones. These flavonoids are ineffective also in the inhibition of lipid autoxidation.

16-3-4. *Catechins.* This class of flavonoids is characterized by a single hydroxyl group on the C_3 portion, at the 3-position. *Catechins* are therefore polyhydroxyflavan-3-ols. The substance that gave its name to the whole group, catechin, was first isolated from "catechu", a tanning extract prepared from heartwood of *Acacia catechu.* It occurs in nature in two stereoisomeric forms (+) – *catechin* and (—) – *epicatechin.*

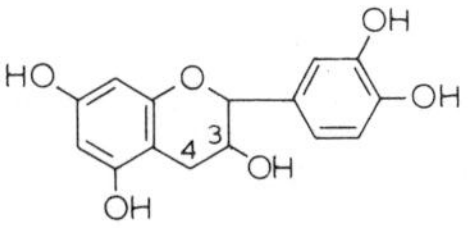

catechin

The two stereoisomers differ in the relative direction of H and OH about the carbon at 3-position. *Gallocatechin*, present in the bark of oak, acacia, chestnut and other trees, has an additional hydroxyl of the 5′ portion.

Catechins take part in the enzymic browning process of many foods. Together with the flavan-3,4-diols (leucoanthocyanidins) they constitute the building blocks of the "*condensed tannins*" (see below).

16-3-5. *Tannins.* The use of extracts from various plants in the curing process of hides and skins (*tannage*) is very old. The active substances of these extracts are polyphenolic substances with molecular weights of 500 to 3000, called *tannins.* Vegetable tannins are classified in two groups: *hydrolyzable tannins* are polyesters of sugars with gallic acid (*gallotannins*), or ellegic acid (*ellagitannins*). The sugar component is usually D-glucose. Part or all the five hydroxyl groups of the sugar may be esterified.

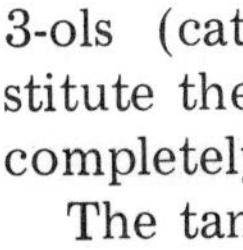

gallic acid

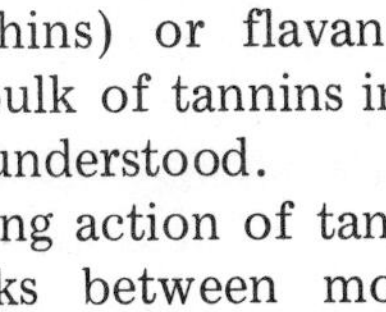

ellagic acid

Condensed tannins are formed by the polymerization of flavan-3-ols (catechins) or flavan-3,4-diols (anthocyanidins). They constitute the bulk of tannins in woods and barks. Their structure is not completely understood.

The tanning action of tannins on hides is based on the formation of cross-links between molecules of *collagen*, especially in the amorphous regions of this protein. These links are believed to involve the hydroxyl groups of thc polyphcnol and the peptide bond of the protein. As a result of tanning, the skins acquire toughness and resistance to bacterial attack and moisture, which are essential properties of good leather. By the same mechanism, tannins denature proteins and precipitate them from their solutions (see Chapter 2).

Tannins are responsible for the *astringency* of many foods, such as apples, pears, persimmons, dates, tea and cocoa. It is generally believed that astringency is associated with protein precipitation since other protein denaturants such as alcohols and heavy metal salts are also astringent.

16-4. GOSSYPOL

Gossypol, the characteristic pigment of cottonseed, was discovered and isolated by Longmore in 1886. Gossypol has a specific polyphenolic structure, not resembling any of the other groups of phenolic substances discussed so far. It has four benzene rings fused in two symmetric naphthalene groups, each half carrying one aldehyde, three hydroxyl, one methyl and one isopropyl radicals.

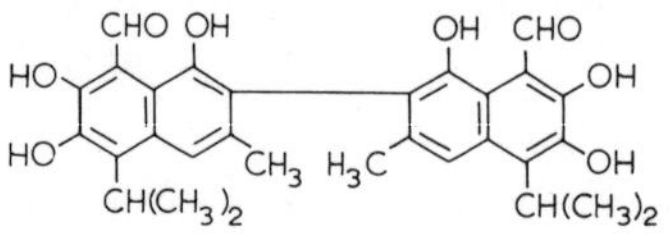

CHO OH OH CHO
HO OH
HO CH_3 H_3C OH
$CH(CH_3)_2$ $CH(CH_3)_2$

gossypol

Gossypol is mainly found in the cotyledons of cottonseed, concentrated in small (100-400 microns) bodies called "pigment glands". It is intensely colored, dark brownish red.

Apart from casual suggestions for its use as a colorant, interest in gossypol has been, for many years, purely academic. However, when cottonseed became an important source of oil, and cottonseed meal was viewed as a potential source of protein for animal and human nutrition, the physiological properties of gossypol emerged as a problem of particular importance. Gossypol is a toxic substance which may cause inflammation of tissues, haemorrhage and nervous disorders. These properties of gossypol exclude the use of untreated cottonseed meal as a food for monogastric animals.

The toxicity of gossypol in cottonseed meal may be eliminated by heating the meal in the presence of moisture. Gossypol, through its carbonylic groups, reacts with the free amino groups of proteins and is converted to "bound gossypol", which is not toxic. However, the nutritional value of cottonseed protein is severely reduced by this treatment, since the first limiting amino acid of cottonseed, lysine, is made unavailable (see Chapter 10).

Nevertheless, heat-detoxified cottonseed meal is used to some extent in animal feeding. It should be noted that the expeller process widely used in the manufacture of cottonseed oil produces a meal with a very low level of "free gossypol", due to the high temperature attained in the steps of cooking and pressing.

Alternative methods have been proposed for the removal of gossypol without damage to the protein. These are based on the extraction of gossypol with specific solvents (aqueous acetone) or on the separation of the whole glands from the rest of the cotyledon, taking advantage of the difference in specific gravity of the two parts of the seed.

A different approach has led to the development of a "glandless cotton" through plant breeding. The seeds of this variety do not contain gossypol.

A large number of pigments with similar structures usually accompany gossypol in cottonseed.

SELECTED BIBLIOGRAPHY TO CHAPTER 16

Geissman, T.A. (Ed.), The Chemistry of Flavonoid Compounds, Pergamon Press, London, 1962.

Haslam, E., Chemistry of Vegetable Tannins, Academic Press, New York, 1966.

Horowitz, R.M. and Gentili, B., Taste and Structure in Phenolic Glycosides, J. Agr. Food. Chem., 17 (1969), 696-700.

Kefford, J.F. and Chandler, B.V., The Chemical Constituents of Citrus Fruits, Academic Press, New York, 1970.

Markman, A.L. and Rzhekhin, V.P., Gossypol and its Derivatives, Israel Program for Scientific Translation, Jerusalem, 1969.

CHAPTER 17

ENZYMIC BROWNING

17-1. INTRODUCTION

The rapid darkening of many fruits and vegetables, such as apples, bananas, avocado, potatoes and eggplants, is a problem with which the food technologist is frequently faced. In contrast with the various classes of non-enzymic browning discussed in Chapter 10, this type of discoloration is very rapid, requires contact of the tissue with oxygen, is known to be enzyme catalyzed, and occurs only in vegetal tissue. The phenomenon is termed "*enzymic browning*". Most frequently, enzymic browning is viewed as an objectionable deterioration process which should be prevented, e.g. in the production of banana puree, potato chips or apple sauce. On the other hand, enzymic browning is essential for the development of the desirable color and flavor in tea and cocoa. Enzymic browning is related to the in vivo synthesis of dark melanin pigments in skin and hair.

The recognition of the enzymic nature of this type of browning in certain fruits should probably be ascribed to Lindet as far back as 1895. However, it was Onslow who showed, in 1920, that the enzymic darkening of plant tissues in air is due to the presence of *o*-dihydroxyphenol derivatives, such as catechol, protocatechuic acid, and caffeic acid as well as hydroxy-gallic acid esters of caffeic acid, such as chlorogenic acid, which is widely distributed in many fruits and especially in potatoes and sweet potatoes.

17-2. MECHANISM

17-2-1. *Overall course.* The initial step of enzymic browning is the enzyme-catalyzed oxidation of *catechol* derivatives to corresponding O-*quinones*. The enzymes involved are termed *polyphenoloxidases*, *catecholases* or *polyphenolases*.

OH, OH, R (catechol derivative) —O, polyphenoloxidase (catecholase)→ O, =O, R (o-quinone)

Quite often, the original substrate is not a polyphenol but rather a monophenol such as the amino-acid *tyrosine*. In this case, the substrate is first hydroxylated to the corresponding *o*-diphenol (catechol). The enzyme involved in this reaction is a *monophenolase* or *cresolase*.

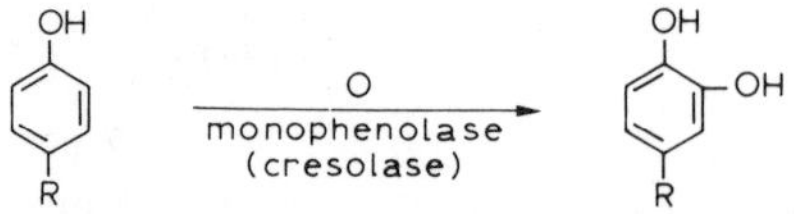

O
monophenolase
(cresolase)

Cresolase activity is illustrated below for the hydroxylation of tyrosine into *3-4-dihydroxyphenylalanine* (DOPA). This reaction is of importance in the enzymic browning of potato tissue and in the biosynthesis of melanin in animals.

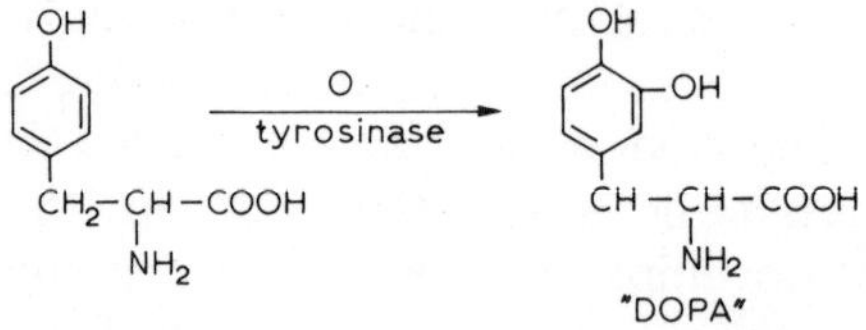

$CH_2-CH(NH_2)-COOH$ —(O, tyrosinase)→ $CH-CH(NH_2)-COOH$

"DOPA"

In enzyme preparations obtained from plant tissue, it is difficult to separate the two types of phenolase activity. Early workers even questioned the existence of cresolase activity and postulated the non-enzymic oxidation of monophenols by *o*-quinones in the following manner:

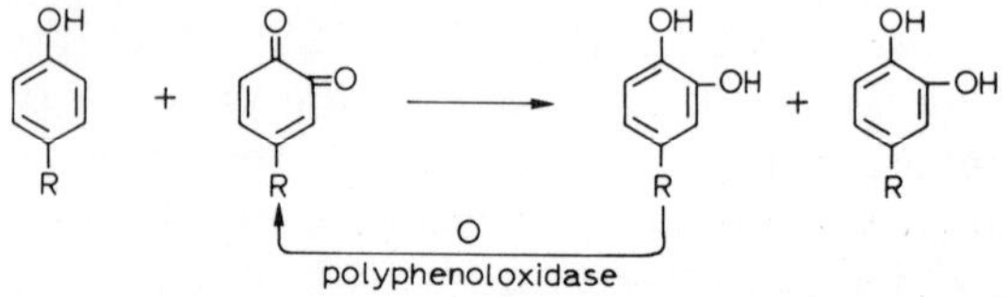

O
polyphenoloxidase

The next step involves the polymerization of *o*-quinones into colored, complex substances. The exact structure of these pigments is not known. The polymerization of *o*-quinones is thought to be preceded by hydroxylation to corresponding hydroxyquinones.

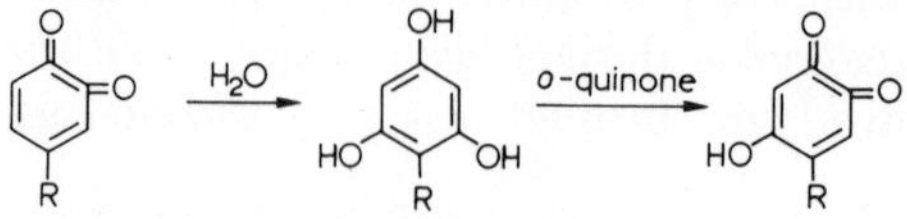

H_2O
o-quinone

This and the subsequent polymerization or condensation reactions leading to red-purple-brown-black pigments are apparently non-enzymic and do not require the presence of oxygen.

17-2-2. *Polyphenoloxidases.* The isolation and characterization of polyphenoloxidases is made difficult by the simultaneous presence of the enzymes and their polyphenolic substrates in plant tissue homogenates. The products of the oxidative reaction combine with the enzyme to form inactive complexes. In order to overcome this difficulty, the tissue is treated with cold acetone. This solvent dehydrates and precipitates the proteins while extracting the interfering phenolic substances. The precipitate known as the "*acetone powder*" is a crude concentrate from which the enzymes may be extracted for further purification.

Polyphenolases are copper-proteins. In the active enzyme the copper is monovalent. Molecular weights in the range of 25-35,000 have been reported. Usually, the polyphenolase system from a given source is not homogeneous but consists of a mixture of a number of enzymes. The optimum pH is near 7. The enzymes are relatively resistant to heat. Depending on the source, as much as 2 to 10 minutes are required in order to inactivate the enzyme at 100° C.

Cresolase is an oxygen-transferase as it catalyzes the transfer of oxygen to the substrate. On the other hand, the oxidation mechanism of catecholase resembles that of a dehydrogenase as it removes hydrogen from the substrate.

17-2-3. *Substrate systems.* In the presence of combined cresolase and catecholase activity, both monophenols and polyphenols may serve as starting materials for enzymic browning. However, in view of their concentration and reactivity some phenolic substances are particularly important. Thus, *chlorogenic acid* (see Chapter 16) and its derivatives are the main substrates of enzymic browning in deciduous fruits.

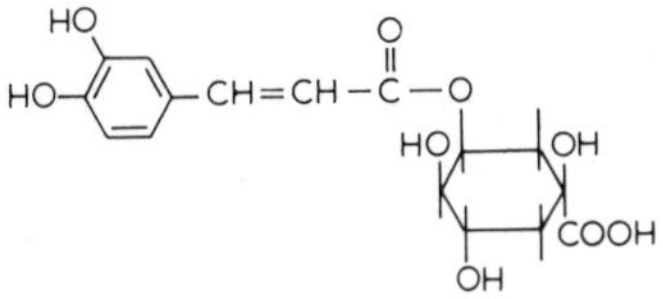

chlorogenic acid

In tea, the predominant substrates are catechins. While the exact chemistry of the final dark pigments is not known, the structure of some intermediate polymers has been elucidated. One of these, *theaflavin*, is an orange colored oxidized dimer of tea catechins.

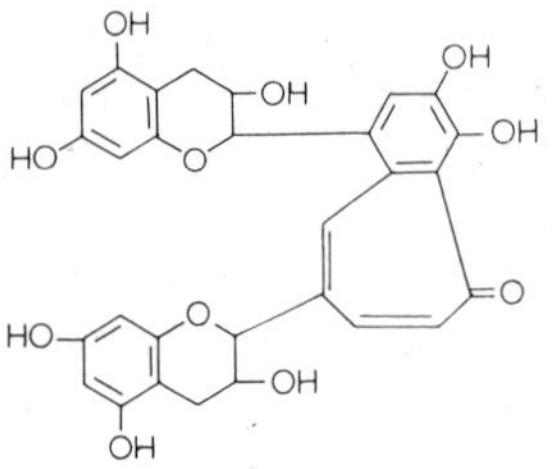

theaflavin

The carbon-to-carbon condensation of the quinone group with a phenolic ring is apparently the universal pattern of polymerization.

The pigment forming substance of potatoes is the amino acid tyrosine. In bananas, the main browning substrate is also a nitrogen-containing phenol : *3,4-dihydroxyphenylethylamine.*

In animals, tyrosine is the precursor of *melanin*, the dark pigment of skin and hair.

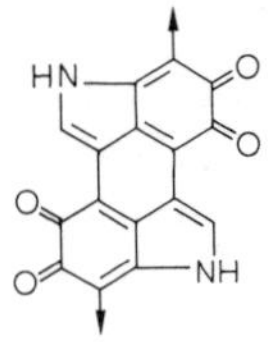

portion of melanin

Amines, amino acids and similar nitrogen-containing compounds react with *o*-quinones to give intensely colored complexes. This interaction is easily demonstrated in vitro and may be of importance in enzymic pigmentation of foods. In some foods, such as cocoa, the addition of amines has been shown to enhance pigment formation.

Under certain conditions, a pink coloration of white onions, garlic and leeks is observed. This phenomenon, highly undesirable in the dehydrated vegetable industry, is closely related to enzymic browning. Here, amino acids seem to be essential reactants. The process apparently starts with the enzymic oxidation of phenolic substances to corresponding quinones, followed by the non-enzymic condensation of these quinones with amino acids.

17-3. CONTROL OF ENZYMIC BROWNING

For enzymic browning to take place, three factors must be present: appropriate phenolic substrates, active phenoloxidases and oxygen. In the practical control of enzymic browning, little can be done in the way of eliminating the substrate. Several methods for the removal or modification of the substrates have been tested but none are in practical use. These include in situ biochemical transformation of the substrates into derivatives less susceptible to browning. Observing that methyl ethers of *o*-diphenols are much more resistant to oxidation than the parent phenols, Finkle proposed a method for preventing browning, based on in situ methylation of natural polyphenols by means of the enzyme *O-methyltransferase* as follows:

$$\text{(OH, OH, }R_1\text{-substituted benzene)} + R_2CH_3 \xrightarrow[\text{transferase}]{\text{O-methyl-}} \text{(OH, }OCH_3\text{, }R_1\text{-substituted benzene)} + R_2H$$

In this way, chlorogenic acid, for instance, is transformed into 3-feruloylquinic acid. Further oxidation to *o*-quinones is thus blocked and browning is prevented. In the practical method proposed, the fruit or vegetable is macerated in a solution containing *o*-methyl transferase and a suitable methyl donor (such as a derivative of methionine) at slightly alkaline pH.

The methods of prevention most commonly practised concentrate on the enzyme and oxygen. The principal approaches are:

(a) To inactivate the enzyme (blanching, inhibitors).

(b) To render the conditions unfavorable to enzyme action (lowering the pH, reduction of water activity).

(c) To minimize contact with oxygen.

(d) To use antioxidants (ascorbic acid, sulfur dioxide).

17-3-1. *Thermal inactivation of phenolases.* The most commonly used method is blanching, i.e. heating the fruit or vegetable in hot water or steam until the enzyme is inactivated. Fruits that will be pulped subsequently may be blanched continuously in so-called fruit scalders, which consist of a screw conveyor where the fruit is heated in live steam. This is the method commonly used in the manufacture of apple sauce and banana puree. However, this method has several limitations. In the first place, it is not applicable to fruits and vegetables which are not to be cooked for reasons of flavor or texture (avocado, potato chips). Furthermore, blanching disrupts the cellular structure and enhances the contact between the enzyme and the substrate. If the thermal treatment is not complete and sufficiently rapid, this may accelerate browning instead of preventing it. Thus, "pinking" is more serious in lightly blanched leeks and onions than in unblanched material. The use of microwave energy as a means of more rapid supply of heat has been suggested.

17-3-2. *Use of acids.* The optimum pH for most phenolases in foods is in the vicinity of 7. Lowering the pH below 4 retards phenolase activity considerably. This is a method widely used in kitchen and industry. Citric acid is most commonly used. Part of its action may be due to its chelating effect on copper (see Chapter 12). The methods of application include dipping the fruit or vegetable in dilute solutions of citric acid or direct addition of the acid to purees, syrups, etc. Malic acid is even more effective.

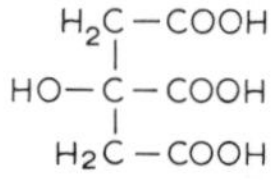

$$\begin{array}{c} H_2C-COOH \\ | \\ HO-C-COOH \\ | \\ H_2C-COOH \end{array} \qquad \begin{array}{c} H_2C-COOH \\ | \\ HO-CH-COOH \end{array}$$

citric acid 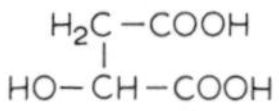 — malic acid

17-3-3. *Ascorbic acid.* Ascorbic acid (or its isomer, iso-ascorbic acid) retards enzymic browning in virtue of its reducing power. It reduces *o*-quinones back to the parent *o*-diphenols.

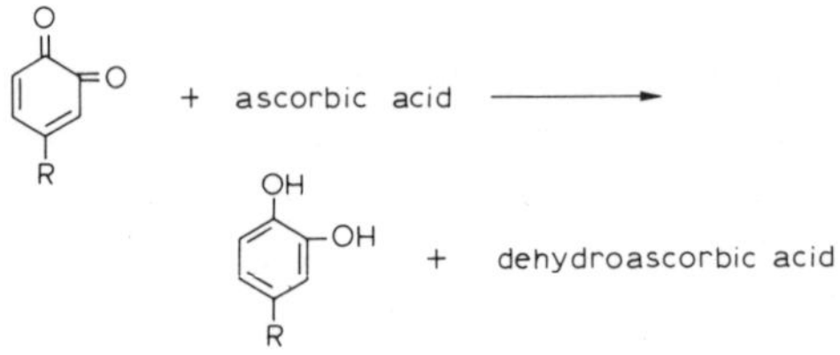

$$\text{o-quinone (R)} + \text{ascorbic acid} \longrightarrow \text{o-diphenol (OH, OH, R)} + \text{dehydroascorbic acid}$$

Ascorbic acid, mostly in combination with citric acid, is widely used as a browning inhibitor in the frozen fruit industry. The time honored culinary use of lemon juice for the same effect is also based on the combined action of ascorbic acid and citric acid. It should be remembered that ascorbic acid itself undergoes browning (see Chapter 10), but this is a slow, non-enzymic process of little consequence in frozen fruit.

Other reducing substances, such as compounds bearing —SH groups, are known to prevent enzymic browning by reducing *o*-quinones. A suitable sulfhydryl-bearing additive which would meet the requirements of safety, organoleptic acceptability, effectiveness and low cost has not been yet proposed.

17-3-4. *Sulfur dioxide.* Sulfur dioxide, sulfites and bisulfites are effective browning inhibitors. They have the advantage of preventing both enzymic and non-enzymic browning, at the same time inhibiting fermentation.

They are effective at very low concentrations, such as a few parts per million of free SO_2. Sulfur dioxide and sulfites are widely applied in the processed fruit industry and for the preservation of peeled, sliced potatoes for institutional use. The mechanism of action is not well understood. Sulfur dioxide is known to inhibit phenoloxidase systems. On the other hand, being a powerful reducing substance, sulfur dioxide may also react by reducing *o*-quinones, just as ascorbic acid. A third mechanism proposed by Embs and Markakis involves the addition of sulfur dioxide to *o*-quinones and prevention of the subsequent polymerization into dark pigments. Probably all three mechanisms are important in the inhibition of enzymic browning by SO_2.

Sulfur dioxide and the salts of sulfurous acid have several limitations which must be borne in mind. They bleach anthocyanin pigments; they destroy thiamine (vitamin B_1); their odor and taste are objectionable at concentrations over 30-50 ppm and the laws in many countries limit their use to specific applications only.

17-3-5. *Use of salt.* Immersion of vegetables in dilute solutions of sodium chloride immediately after peeling and slicing has been long known to retard browning. This effect is believed to be due to the inactivation of the phenolase system by salt. Salt concentrations of a few grams per liter are sufficient. For reasons of taste, the method is limited to vegetables and not easily applicable to fruit.

17-3-6. *Prevention of contact with oxygen.* This is an efficient method for the prevention of oxidative browning. However, due to technical difficulties in its implementation, it is used only when other methods are not applicable. A typical case of this class is the prevention of oxidative browning in avocado products. Thermal inactivation of enzymes cannot be used because an intensely bitter taste develops on the slightest heating. The pH cannot be lowered too far for reasons of flavor. Ascorbic acid is only partially effective. Even when acidified to the level of organoleptic acceptability and in the presence of ascorbic acid, refrigerated avocado products darken at the free surface. Packaging under nitrogen prevents surface discoloration in this case.

SELECTED BIBLIOGRAPHY TO CHAPTER 17

Course, J., The Enzymatic Browning of Fruits and Vegetables, in Phenolics in Normal and Diseased Fruits and Vegetables, V.C. Runeckles (Ed.), Norwood, Mass., 1964.

Embs, J. and Markakis, P., The Mechanism of Sulfite Inhibition of Browning Caused by Polyphenol Oxidase, J. Food Sci., **30** (1965), 753.

Finkle, B.J., Treatment of Plant Tissue to Prevent Browning, U.S. Patent 3 (1964), 126, 287.

Joslyn, M.A. and Ponting, J.D., Enzyme-catalyzed Oxidative Browning of Fruit Products, Adv. Food Res., **3** (1951), 1.

Mathew, A.G. and Parpia, H.A.H., Food Browning as Polyphenol Reaction, Adv. Food Res., **19** (1971), 75.

Palmer, J.K., Banana Polyphenoloxidase; Preparation and Properties, Plant Physiol., **38** (1963), 508.

CHAPTER 18

RESPIRATION

Respiration is combustion

ANTOINE LAVOISIER, 1789

18-1. INTRODUCTION

Modern biochemistry defines respiration as that process of Life by which electrons are transferred from organic substrates (fuels) to molecular oxygen. By this definition, Lavoisier's concept of respiration appears to be remarkably accurate. To the biochemist, respiration is important as the mechanism whereby aerobic cells release most of the chemical energy available in the "fuel" substrates. In food biochemistry, the subject is of particular interest in connection with nutritional biochemistry and with the post-harvest behavior of fruits and vegetables. The biochemistry of respiration also explains the origin of many intermediate metabolites of importance in foods, such as plant acids.

The two metabolic processes in the living cell (in animals, plants and in microorganisms) which are engaged in exploiting the energy contained in its food are *fermentation* and *respiration.* While fermentation (Chapter 19) results in products still rich in energy, respiration results in total breakdown of the carbohydrate into the simple oxides, CO_2 and H_2O, of very low potential energy. The substrates for this breakdown under aerobic conditions are simple hexoses, or other organic compounds, acids, fats, etc. The overall reaction of respiration for hexoses can be summarized as follows:

$$C_6H_{12}O_6 + 6O_2 \rightarrow 6CO_2 + 6H_2O \qquad \Delta F^\circ = -686\,\text{Kcal.}$$

Here, the "*respiratory quotient*" (R.Q.), i.e. the ratio of the number of moles of CO_2 produced to the number of moles of O_2 taken up, will be equal to 1:

$$\text{R.Q.} = \frac{6CO_2}{6O_2} = 1$$

In the case of the oxidation of malic acid, for instance, which is richer in oxygen than an hexose, the R.Q. will be larger than 1:

$$C_4H_6O_5 + 3O_2 \rightarrow 4CO_2 + 3H_2O$$

$$R.Q. = \frac{4CO_2}{3O_2} = 1.33$$

In the case of lipids, which are poorer in oxygen, the R.Q. is less than 1. By convention, the value for mixed fats is assumed to be 0.7.

The catabolism of biological fuel starts with the breakdown of storage material to smaller units. Polysaccharides are degraded to single hexoses or pentoses; proteins are broken down to constituent amino acids; fats are hydrolyzed to glycerol and individual fatty acids. All along this first stage, changes in the redox potential of the substrate are nil and the free energy interchange is very small. In the next stage, the simple substrates are further degraded down to three or two carbon units. The first stage and most of the reactions of the second stage are common to aerobic respiration and anaerobic fermentation. Aplied to sugars, which are the predominant biological fuels, the second stage is termed *glycolysis* (Section 18-2). The third stage, characteristic to respiration only, involves the final oxidation of the organic matter to carbon dioxide and water, through a cyclic pathway known as the Krebs tricarboxylic acid cycle (Section 18-3). Most of the energy is released during this last stage. An alternative route of hexose degradation known as the hexose monophosphate shunt or the phosphogluconate pathway is of considerable importance (Section 18-4). The transport of electrons from each particular fuel substrate to molecular oxygen is a process involving several steps and as many enzymes (Section 18-5). The technological aspects of post-harvest respiration of fruits and vegetables are discussed in Section 18-6.

18-2. GLYCOLYSIS

Historically, the elucidation of the mechanism of glycolysis is closely associated with fermentation research. The recognition of at least part of the glycolytic pathway as an initial step to respiration is relatively recent.

The breakdown of simple sugars into alcohol has probably been known since prehistoric times. Ancient Egyptians were well acquainted with the art of fermenting fruit juices into alcoholic beverages and were even able to preserve them from further spoilage in golden jars. In 1810 Gay-Lussac showed that the reaction of alcoholic fermentation can be described as follows:

$$C_6H_{12}O_6 = 2C_2H_5OH + 2CO_2\uparrow$$

Fifty years later (1861) Pasteur demonstrated that fermentation caused by living yeast cells is an anaerobic process, i.e. does not require the participation of atmospheric oxygen; in other words, that by means of this process yeast organisms are able to draw their required energy from glucose in the absence of oxygen, in his own words: "la vie sans air".

In contrast to the other important catabolic process, respiration, during which glucose undergoes complete breakdown in the presence of oxygen into CO_2 and H_2O, products with very low potential energy, fermentation converts glucose into alcohol which, in itself, still contains quite a large proportion of energy.

The above description of fermentation given by Gay-Lussac describes only the overall reaction but it does not show, however, the true biochemical mechanism. The first most important step towards the clarification of this mechanism was made by the Buchners (1897), who found that fermentation can also be achieved without the presence of living yeast cells. Buchners succeeded in causing the breakdown of glucose into ethanol by using a cell-free extract of the yeast, thereby showing for the first time that such metabolic processes are carried out and catalyzed by ferments or enzymes.

While CO_2 and ethanol are the principal products of alcoholic fermentation, it was soon discovered that a number of other products are present during this process (although in very small amounts), such as acetaldehyde (CH_3CHO), glycerol ($CH_2OH.CH.OH.CH_2.OH$), pyruvic acid ($CH_3CO.COOH$), and others. It took quite a long time and involved the participation of a large number of well-known investigators, among them Robison, Cori, Neuberg, Harden and Young, before a final scheme was drawn up by Meyerhof and Embden, which is now universally accepted.

According to this scheme (Fig. 27), the first step is the phosphorylation of *D*-glucose to *D*-glucose 6-phosphate, at the expense of ATP.

This step is catalyzed by the enzyme *hexokinase* (or in some cases

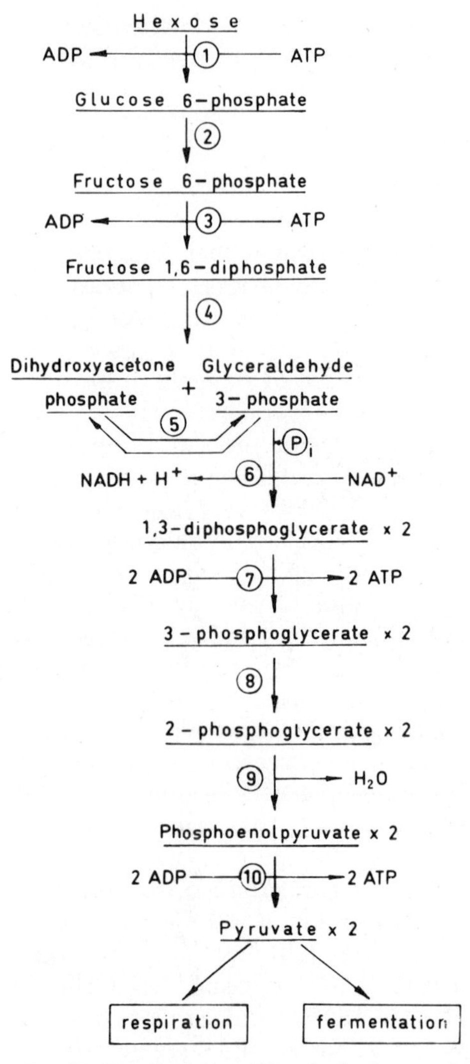

Fig. 27. The Embden–Meyerhof scheme of glycolysis.

by another enzyme, *glucokinase*). Magnesium cations are required, since the true substrate is probably the magnesium salt of ATP.

$$\text{(Step 1)}\quad \text{D-glucose + ATP} \xrightarrow[\text{Mg}^{++}]{\text{Hexokinase}} \text{D-glucose-6-phosphate + ADP}$$

Since a low-energy phosphate bond is formed at the expense of the energy-rich phosphate of ATP, the reaction is strongly exergonic ($\Delta F^0 = -4.0$ kcal).

The second stage is the isomerization of glucose-6-phosphate to *fructose 6-phosphate* by means of the enzyme *phosphoglucoisomerase*, with little change in free energy (Step 2).

At the third stage, fructose-6-phosphate is further phosphorylated to *fructose 1-6-diphosphate*, at the expense of an additional ATP molecule. The reaction is catalyzed by the enzyme *phosphofructokinase.*

$$\text{(Step 3)}\quad \text{Fructose-6-phosphate + ADP} \xrightarrow[\;]{\text{Phosphofructokinase}\;Mg^{++}} \text{Fructose 1-6-diphosphate + ADP}$$

Up to this point the carbon skeleton of the hexose remains intact. The purpose of these initial stages is to activate or "prime" the hexose molecule for the subsequent process of cleavage.

The next step is the fission of the diphosphorylated hexose to two phosphotrioses: *glyceraldehyde-3-phosphate* and *dihydroxyacetone phosphate.* The reaction which is essentially an aldol condensation in reverse, is catalyzed by the enzyme *aldolase.*

$$\text{(Step 4)}\quad \text{Fructose 1-6-diphosphate} + H_2O \xrightarrow{\text{Aldolase}}$$

$$\longrightarrow \begin{array}{c} HCO \\ | \\ HCOH \\ | \\ H_2C-O-(P) \end{array} \quad + \quad \begin{array}{c} H_2C-OH \\ | \\ C{=}O \\ | \\ H_2C-O-(P) \end{array}$$

glyceraldehyde-3-phospate dihydroxyacetone phosphate

The two isomeric trioses are interconvertible. The interconversion is catalyzed by the enzyme *triosephosphate isomerase* (Step 5).

Now comes a step of particular importance in which the oxidation level of the substrate is altered for the first time. This is the oxidation of the carbonyl group in glyceraldehyde-3-phosphate to a carboxyl group.

The resulting acid is simultaneously phosphorylated at the expense of inorganic phosphate and the end result is *1-3-diphosphoglyceric acid.* The oxidizing enzyme is *glyceraldehyde-3-phosphate dehydrogenase* requiring NAD.

(Step 6)

$$\begin{array}{c} \mathrm{HCO} \\ | \\ \mathrm{HCOH} \\ | \\ \mathrm{H_2C{-}O{-}(P)} \end{array} + \mathrm{NAD^+} + \mathrm{(P)_i} \longrightarrow \begin{array}{c} \mathrm{O} \\ \| \\ \mathrm{C{-}O{\sim}(P)} \\ | \\ \mathrm{HCOH} \\ | \\ \mathrm{H_2C{-}O{-}(P)} \end{array} + \mathrm{NADH} + \mathrm{H^+}$$

The energy released by oxidation is captured in the high energy phosphate bond of 1-3-diphosphoglyceric acid. In the next step this bond is transferred to ADP, to regenerate a molecule of ATP.

(Step 7)

$$\text{1-3-diphosphoglyceric acid} + \text{ADP} \xrightarrow{\text{phosphoglycerate kinase}} \text{3-phosphoglyceric acid} + \text{ATP}$$

The *3-phosphoglyceric* acid is now isomerized by the enzyme *phosphoglyceromutase* into *2-phosphoglyceric acid* (Step 8) and then by losing a molecule of water, converted into *phosphoenolpyruvic acid.*

(Step 9)

$$\begin{array}{c} \mathrm{COOH} \\ | \\ \mathrm{C{-}O{-}(P)} \\ | \\ \mathrm{H_2C{-}OH} \end{array} \xrightarrow{-\mathrm{H_2O}} \begin{array}{c} \mathrm{COOH} \\ | \\ \mathrm{C{-}O{\sim}(P)} \\ \| \\ \mathrm{CH_2} \end{array}$$

In this last reaction, catalyzed by the enzyme *enolase*, the phosphate ester link becomes a high-energy phosphate bond, although the overall standard free energy change of the reaction is very small.

Phosphoenolpyruvate now loses its phosphoric acid to ADP in a strongly exergonic reaction catalyzed by *pyruvic kinase.*

(Step 10)

$$\begin{array}{c} \mathrm{COOH} \\ | \\ \mathrm{C{-}O{\sim}(P)} \\ \| \\ \mathrm{CH_2} \end{array} \xrightarrow[\mathrm{ADP}\ \ \mathrm{ATP}]{} \left[\begin{array}{c} \mathrm{COOH} \\ | \\ \mathrm{C{-}OH} \\ \| \\ \mathrm{CH_2} \end{array}\right] \rightleftharpoons \begin{array}{c} \mathrm{COOH} \\ | \\ \mathrm{C{=}O} \\ | \\ \mathrm{CH_3} \end{array}$$

pyruvic acid

Pyruvic acid may be considered as the pivot of the catabolic pathway. Up to this stage, the glycolytic sequence is common to aerobic and anaerobic processes. The fate of pyruvate, however, now depends on the type of catabolism. In the case of respiration, it is oxidized and decarboxylated to *acetate* (Section 18-3). In anaerobic processes it is reduced to *lactic acid*, by the action of lactate dehydrogenase.

$$\begin{array}{c} \mathrm{COOH} \\ | \\ \mathrm{C{=}O} \\ | \\ \mathrm{CH_3} \end{array} + \mathrm{H^+} + \mathrm{NADH} \longrightarrow \begin{array}{c} \mathrm{COOH} \\ | \\ \mathrm{HC{-}OH} \\ | \\ \mathrm{CH_3} \end{array} + \mathrm{NAD^+}$$

lactic acid

The anaerobic process is characterized by the absence of an external electron acceptor such as molecular oxygen. The reduction of pyruvate is therefore essential for the regeneration of the oxidative power, NAD^+, lost at the stage of oxidation of phosphoglyceraldehyde to phosphoglyceric acid (Step 6). Lactic acid is the last station of anaerobic glycolysis in the muscle (see Section 5-1) or in the lactic acid fermentation (Chapter 19). Pyruvate or lactate may enter a series of other reactions at the end of which other fermentation products such as *ethanol* or *glycerol* are formed (see Chapter 19).

If the sequence of anaerobic glycolysis from glucose to lactic acid is considered, the net result is:

$$\underset{\text{Glucose}}{C_6H_{12}O_6} \rightarrow \underset{\text{Lactic acid}}{2C_3H_6O_3}$$

There is no net change in oxidation state but only internal oxidation–reduction of the substrate. Energy is released at the rate of some 36 kcal/mol of glucose, most of which is stored in the form of the energy-rich phosphate bond of two ATP molecules (two ATP's are used up at the initial phosphorylation steps 1 and 3, but two are generated at step 7 and two at step 10).

18-3. THE TRICARBOXYLIC ACID OR KREBS' CYCLE

In the respiration process, pyruvic acid is oxidized and degraded in a series of reactions in cyclic sequence. The cycle known as the *tricarboxylic acid* or the *Krebs' cycle* was proposed by Hans Krebs in 1937, in the light of some established isolated facts and his own experimental work. The cyclic process, as shown in Fig. 28, is now accepted as the universal pathway of respiration.

The overall reaction of the degradation of pyruvate may be represented as follows:

$$CH_3\ CO\ COOH + 3H_2O \rightarrow 3CO_2 + 10H$$

As can be seen, the significant elements of this reaction consist of *hydration* (addition of water), *decarboxylation* (removal of CO_2) and *dehydrogenation.* Molecular oxygen does not enter the process directly, but serves as the terminal acceptor of hydrogen, so that the end products are really CO_2, water and, of course, energy. Several enzymes and co-enzymes catalyze these processes. Let us first review some of the coenzymes before the description of the cycle.

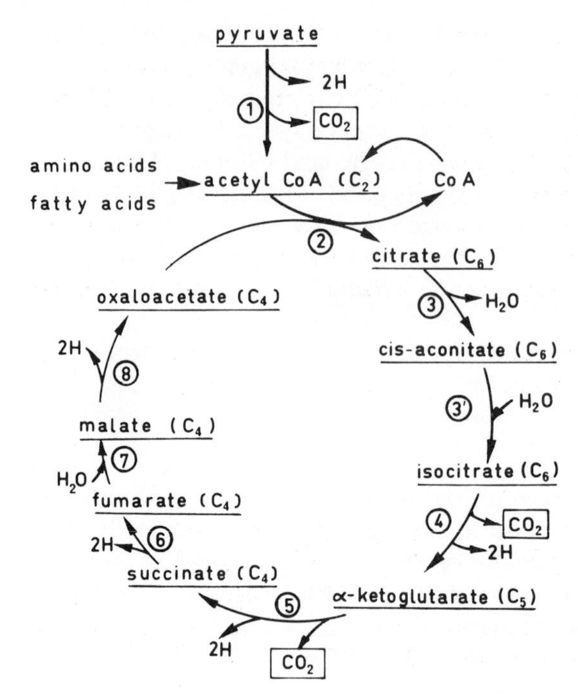

Fig. 28. Tricarboxylic acid or Krebs' cycle.

Dehydrogenation processes are catalyzed by specific dehydrogenases which transfer hydrogen from their substrates to NAD, NADH or FAD. These nucleotide coenzymes and their relation to vitamins were considered in Section 3-3.

Decarboxylation is catalyzed by decarboxylase in the presence of the coenzyme *thiamine pyrophosphate* (TPP) or *cocarboxylase*. The essential part of the cocarboxylase molecule is *thiamine*, or vitamin B_1.

thiamine pyrophosphate (cocarboxylase)

Before entering the Krebs' cycle, pyruvate is decarboxylated and dehydrogenated to acetate. The coenzyme responsible for the transfer of acetate into the cycle is *Coenzyme A*, discovered by Lipmann and his group in 1950.

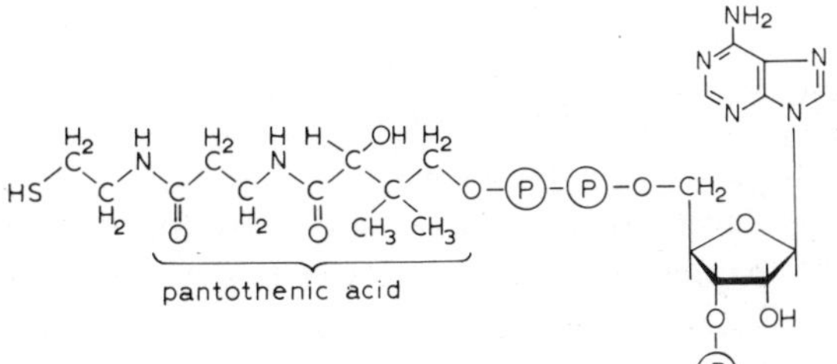

pantothenic acid

coenzyme A

One of the components of the CoA molecule, *pantothenic acid*, is also a vitamin of the vitamin B group.

The most important property of Coenzyme A is its ability to combine with acetate and to transfer it to appropriate substrates. The Co-A-acetate complex, *acetyl-CoA*, is referred to as "*active acetate*". In this compound, acetyl is bound to the sulfur atom through an energy-rich bond:

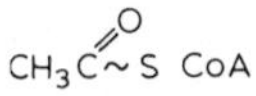

$$CH_3\overset{O}{\overset{/\!/}{C}}\sim S\ CoA$$

acetyl CoA (active acetate)

Now, the individual steps of the Krebs' cycle will be briefly described.

(1) In the first step, pyruvate is oxidized and decarboxylated to active acetate. The reaction is complex and requires three enzymes and five co-enzymes, among which are NAD, CoA, TPP. The overall reaction is:

$$CH_3COCOO^- + CoA + NAD^+ \rightarrow \text{Acetyl CoA} + CO_2 + NADH + H^+$$

(2) The second step is the transfer of acetyl from acetyl CoA to *oxaloacetate* (from the cycle) to form *citric acid*. The reaction is catalyzed by the enzyme *citrate synthase* (formerly known as *condensing enzyme*).

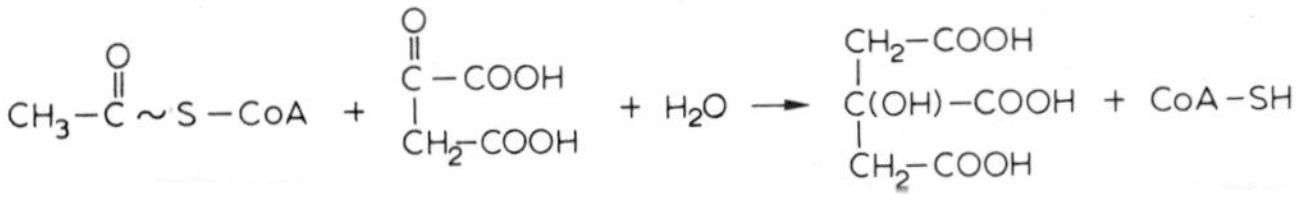

$$CH_3-\overset{O}{\overset{\|}{C}}\sim S-CoA + \begin{matrix}\overset{O}{\overset{\|}{C}}-COOH \\ | \\ CH_2-COOH\end{matrix} + H_2O \longrightarrow \begin{matrix}CH_2-COOH \\ | \\ C(OH)-COOH \\ | \\ CH_2-COOH\end{matrix} + CoA-SH$$

acetyl CoA | oxalcacetic acid | citric acid

(3) The citric acid formed in this way is held in equilibrium with its isomer *isocitric acid* and *aconitric acid* through the enzyme *aconitase*. Aconitase requires bivalent iron.

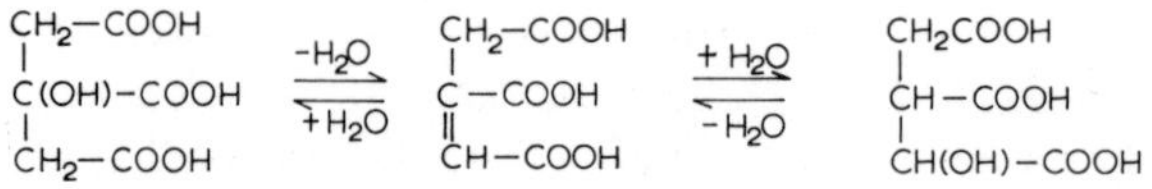

$$\begin{array}{c} CH_2-COOH \\ | \\ C(OH)-COOH \\ | \\ CH_2-COOH \end{array} \underset{+H_2O}{\overset{-H_2O}{\rightleftharpoons}} \begin{array}{c} CH_2-COOH \\ | \\ C-COOH \\ \| \\ CH-COOH \end{array} \underset{-H_2O}{\overset{+H_2O}{\rightleftharpoons}} \begin{array}{c} CH_2COOH \\ | \\ CH-COOH \\ | \\ CH(OH)-COOH \end{array}$$

citric acid — aconitic acid — isocitric acid

(4) In the next step, isocitrate is simultaneously dehydrogenated and decarboxylated to *α-ketoglutarate*. The reaction is catalyzed by *isocitrate dehydrogenase* and the hydrogen is transferred to either NAD or NADP.

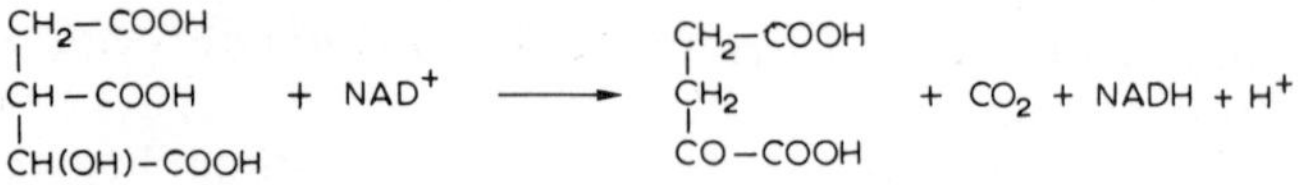

$$\begin{array}{c} CH_2-COOH \\ | \\ CH-COOH \\ | \\ CH(OH)-COOH \end{array} + NAD^+ \longrightarrow \begin{array}{c} CH_2-COOH \\ | \\ CH_2 \\ | \\ CO-COOH \end{array} + CO_2 + NADH + H^+$$

isocitric acid — α-ketoglutaric acid

(5) The next reaction is the oxidative decarboxylation of ketoglutarate to *succinate*. This reaction resembles the transformation of pyruvate to active acetate as it requires TPP, NAD and CoA. It occurs in two stages. Succinate emerges from the first step as the *succinyl-CoA* complex. The decomposition of this complex to succinate and CoA is coupled with the phosphorylation of *guanosine diphosphate* (GDP) to GTP, at the expense of the energy liberated from the high-energy thioester bond. The net result of these two stages is:

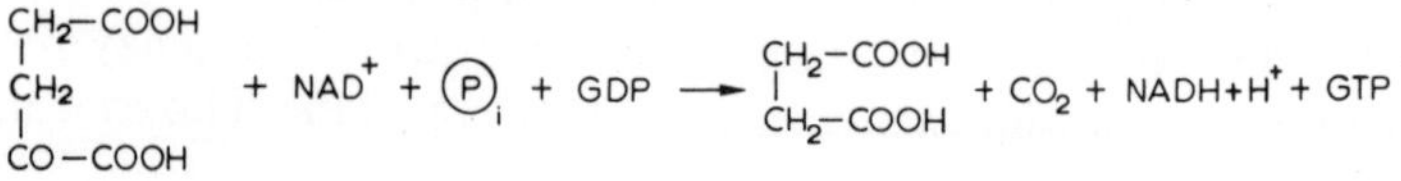

$$\begin{array}{c} CH_2-COOH \\ | \\ CH_2 \\ | \\ CO-COOH \end{array} + NAD^+ + (P)_i + GDP \longrightarrow \begin{array}{c} CH_2-COOH \\ | \\ CH_2-COOH \end{array} + CO_2 + NADH + H^+ + GTP$$

α-ketoglutaric acid — succinic acid

GDP is regenerated through the transfer of phosphate from DGDP to form ATP.

(6) Succinate is now oxidized to *fumarate* by the enzyme *succinate dehydrogenase* containing FAD.

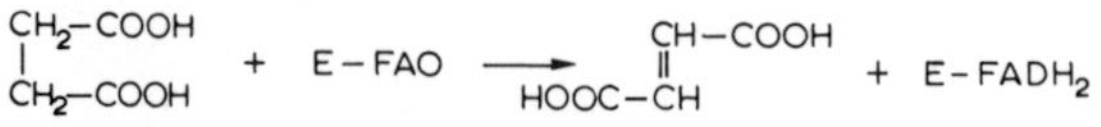

$$\begin{array}{c} CH_2-COOH \\ | \\ CH_2-COOH \end{array} + E-FAO \longrightarrow \begin{array}{c} CH-COOH \\ \| \\ HOOC-CH \end{array} + E-FADH_2$$

succinic acid — fumaric acid

(7) In the next step, fumarate is hydrated to *malate*, this reaction being catalyzed by the enzyme *fumarase*:

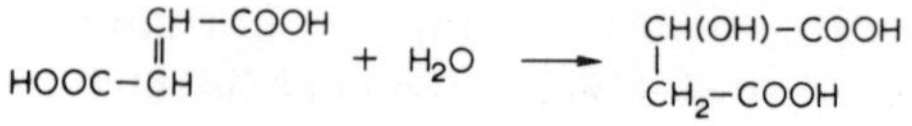

$$\begin{array}{c} CH-COOH \\ \| \\ HOOC-CH \end{array} + H_2O \longrightarrow \begin{array}{c} CH(OH)-COOH \\ | \\ CH_2-COOH \end{array}$$

fumaric acid — malic acid

(8) In the last step of the cycle, malate is dehydrogenated by means of NAD requiring *malate hydrogenase*, to form *oxaloacetic acid* which is now ready to accept a further molecule of acetyl-CoA and to start the cycle anew:

$$\begin{array}{l}CH(OH)-COOH\\ |\\ CH_2-COOH\end{array} + NAD^+ \longrightarrow \begin{array}{l}CO-COOH\\ |\\ CH_2-COOH\end{array} + NADH + H^+$$

malic acid — oxaloacetic acid

The importance of the tricarboxylic acid cycle is not limited to the aerobic catabolism of carbohydrates. This cycle represents the major pathway for the final oxidation of fats and proteins as well.

It will be recalled (Section 11-2) that most natural fatty acids contain an even number of carbon atoms. This fact was long regarded as a clue to the breakdown (or biosynthesis) of fatty acids in steps involving the removal (or addition) of C_2 fragments. This view has been supported by a wealth of experimental evidence. The present theories on fatty acid oxidation recognize the following general pathway:

(a) A fatty acid having 2n carbon atoms is first activated by condensation with CoA.

(b) The fatty acid. CoA complex is oxidized (dehydrogenated) into the corresponding unsaturated homolog with a double bond between carbon atoms (n—3) and (n—2).

(c) The unsaturated acid. CoA complex is hydrated and oxidized so that a keto function now appears in position (n—3) or β.

(d) Now the complex splits into acetyl — CoA and a fatty acid with n—2 carbon atoms. Acetyl–CoA is incorporated in the Krebs' cycle while the degraded fatty acid is again activated, oxidized, shortened by two more carbon atoms and so on. The oxidation of fatty acids in this manner occurs exclusively in the mitochondria.

The oxidative degradation of amino acids is more complex. The degradation process is different for each amino acid and may also differ from one organism to another. At the end, however, all the amino acids are converted to a small number of organic acids, all participants of the Krebs' cycle. (Acetate, pyruvate, oxaloacetate, fumarate, succinate and α-ketoglutarate). At one stage or another, the process requires the removal of the amino group. In the case of most amino acids this is done by *transamination*, i.e. the transfer of the amino group from an amino acid to an α-keto acid, as follows:

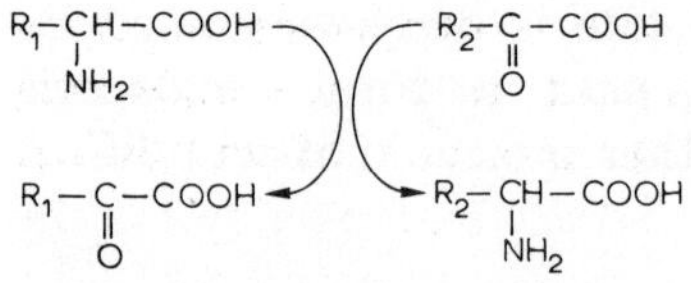

$$R_1-CH(NH_2)-COOH + R_2-CO-COOH \rightleftharpoons R_1-CO-COOH + R_2-CH(NH_2)-COOH$$

This reaction is catalyzed by *transaminases.* Transaminases require the co-enzyme *pyridoxal phosphate*, a derivative of a group of interchangeable dietary factors known as *vitamin B_6*.

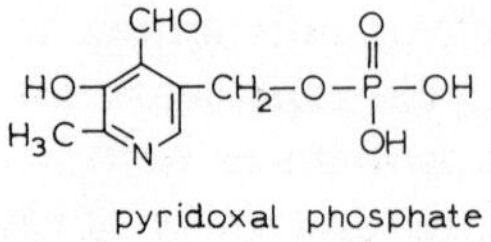

pyridoxal phosphate

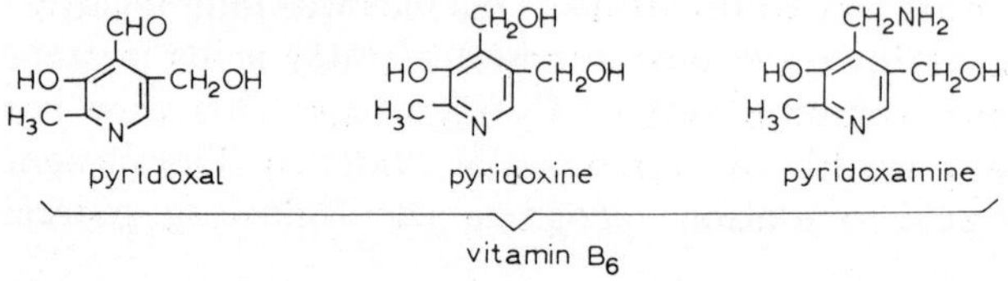

pyridoxal   pyridoxine   pyridoxamine

vitamin B_6

Any of the three α-keto acids of the tricarboxylic acid cycle (pyruvate, oxaloacetate, α-ketoglutarate) may serve as acceptors of the amino group, forming alanine, aspartic and glutamic acid respectively. However, most of the amino nitrogen removed by transamination is finally recovered as *glutamate.* Glutamate undergoes *oxidative deamination*, catalyzed by NAD-requiring *glutamate dehydrogenase*:

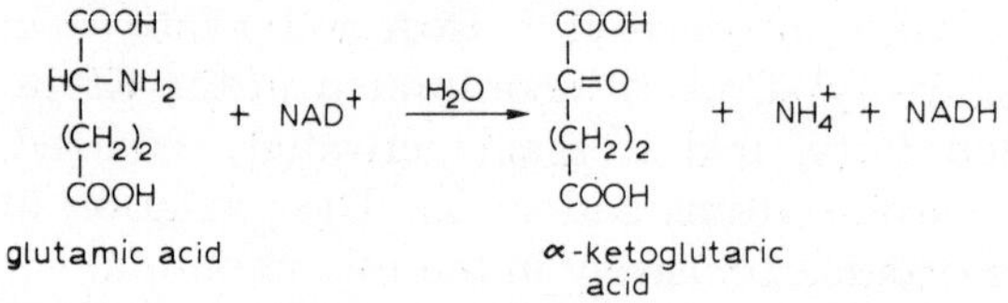

$$HOOC-(CH_2)_2-CH(NH_2)-COOH + NAD^+ \xrightarrow{H_2O} HOOC-(CH_2)_2-CO-COOH + NH_4^+ + NADH$$

glutamic acid   α-ketoglutaric acid

α-Ketoglutaric acid joins the tricarboxylic acid cycle pool. The fate of ammonia depends on the organism. It may be excreted as such (aquatic invertebrates), as trimethylamine oxide (teleost fish), as urea (mammals, sharks), or uric acid (reptiles, birds).

The observation that the catabolic pathway of all three major classes of nutrients — carbohydrates, proteins and fats — converges into the tricarboxylic acid cycle explains the close interrelation between the metabolism of these three groups. Furthermore, the Krebs' cycle often participates in the "upstream" flow of substrates in the metabolic pool, producing the substances needed for anabolic processes (acetyl–CoA for the biosynthesis of fatty acids and some amino acids, α-keto acids as acceptors of NH_2 in the biosynthesis of amino acids, etc.). These anabolic side-streams tend to deplete the medium from the tricarboxylic acid cycle intermediates. To counteract this depletion and keep the cycle in function, an additional mechanism is needed for the synthesis of new oxaloacetate. This mechanism consists in the carboxylation of pyruvate, with a coupled ATP–ADP system supplying the necessary increase in free energy.

In this connection, we meet another water soluble vitamin, *biotin* (sometimes termed *vitamin H*), which serves as a prosthetic group to the enzyme *pyruvic acid carboxylase.* The structure of biotin (which is also important in other carboxylation processes in the biosynthesis of fatty acids) is shown below:

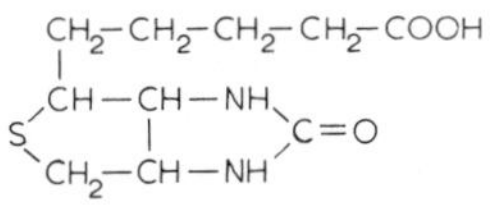

```
 CH2-CH2-CH2-CH2-COOH
 |
 CH—CH—NH
/    |    \
S    |     C=O
\    |    /
 CH2—CH—NH
     biotin
```

18-4. THE HEXOSE MONOPHOSPHATE PATHWAY

The glycolytic pathway, followed by the Krebs' cycle, is probably responsible for 90% of the carbohydrate metabolism during respiration. Nevertheless, it is of importance to mention that additional metabolic pathways have been shown to exist in plants, in some animal tissues and in several types of microorganisms. One of such alternative routes recognized by Horecker in 1953 is the so-called "*pentose-cycle*", also known as the *hexose monophosphate* (HMP) *shunt* or the *phosphogluconate pathway*, which involves the oxidation of glucose 6-phosphate to *6-phosphogluconic acid* and the conversion of the latter into pentose phosphate. This Horecker-Varburg-Dickens route bears considerable likeness to the dark reaction in photosynthesis (Section 20-3), but in reverse.

It has been known for some time that the enzyme *glucose-6-phosphate dehydrogenase*, which contains NADP, is capable of catalyzing the oxidation of D-glucopyranose-6-phosphate into 6-phosphogluconic acid. As a first step gluconolactone is received. This hydrolyzes spontaneously to 6-phosphogluconic acid. The β-keto acid produced after further dehydration is easily decarboxylized. An enzyme, named *6-phosphogluconic acid dehydrogenase*, found in *Escherichia coli*, is capable of catalyzing the oxidative decarboxylation of this acid to form D-ribulose 5-phosphate:

CHO		COOH		COOH		CH_2OH
HCOH		HCOH		HCOH		C=O
HOCH	$\xrightarrow{-2H}$	HOCH	$\xrightarrow{-2H}$	C=O	$\xrightarrow{-CO_2}$	HCOH
HCOH		HCOH		HCOH		HCOH
HCOH		HCOH		HCOH		CH_2O–Ⓟ
CH_2–O–Ⓟ		CH_2–O–Ⓟ		CH_2O–Ⓟ		
glucose 6-phosphate		6-phospho-gluconic acid		6-phospho-3-ketogluconic acid		ribulose 5-phosphate

The pentose so obtained forms an equilibrium between *xylulose-5-phosphate* and *ribose 5-phosphate*. These two pentoses are transformed by means of the enzyme *transketolase* into a triose and

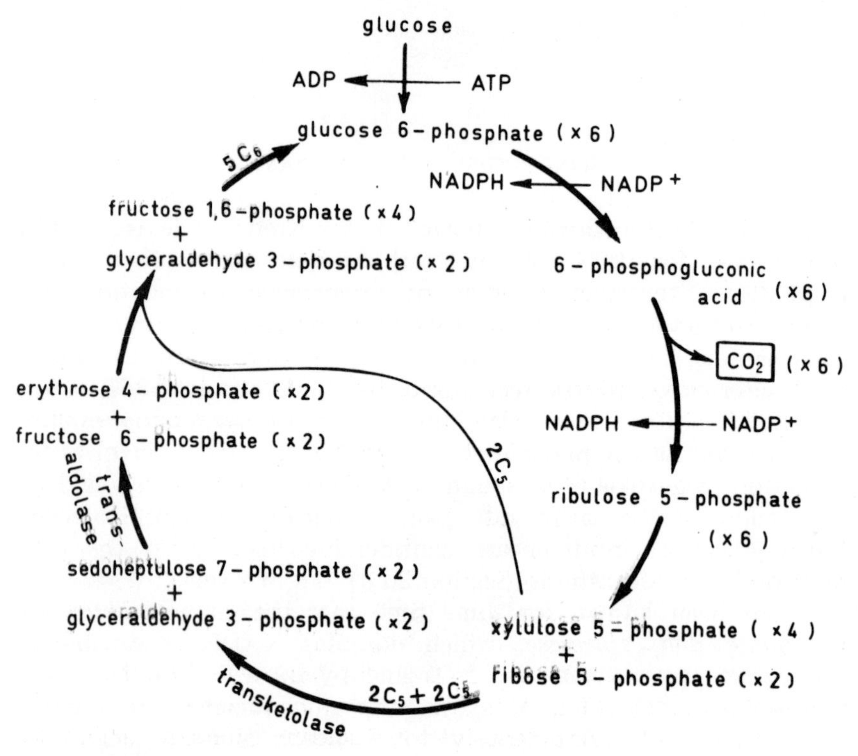

Fig. 29. The pentose oxidation cycle (hexose-monophosphate shunt).

a seven carbon compound, *glyceraldehyde 3-phosphate* and *sedoheptulose 7-phosphate*, respectively. The dihydroxy-acetone portion of the sedoheptulose is easily cleaved by another enzyme, *transaldolase*, and transferred to the glyceraldehyde to form a hexose, while the remaining 4-carbon sugar (*erythrose 4-phosphate*) combines further with a pentose, as shown in Fig. 29.

The overall reaction of hexose oxidation via the pentose cycle a can be summarized as follows:

$$6\,C_6 + 6\,O_2 \longrightarrow 6\,C_5 + 6\,CO_2$$

$$4\,C_5 \longrightarrow 2\,C_6 + 2\,C_4$$

$$2\,C_5 + 2\,C_4 \longrightarrow 2\,C_6 + 2\,C_3$$

$$2\,C_3 \longrightarrow C_6$$

In other words, only one hexose molecule out of every six enters the cycle, is changed into $6CO_2$ and $6H_2O$, the remaining five molecules enter the cycle again.

18-5. THE RESPIRATORY CHAIN

Oxidation of a substrate means the removal of electrons from its molecule. In respiration, molecular oxygen serves as the ultimate acceptor of electrons removed from the oxidized substrates. The direct transfer of a pair of electrons from an organic compound such as malic acid to molecular oxygen would involve a tremendous decrease in free energy. In nature, however, this transfer is carried out in a series of reactions, and the picture is one of a cascade rather than that of a single waterfall. Electrons are transferred from the substrate to an intermediate carrier A, which passes them on to carrier B, etc., until the last carrier of the chain transfers the electrons to molecular oxygen. Each step is enzyme-catalyzed and the carriers are in fact co-enzyme. Of course, the chain must be consistent with the redox potential of the consecutive carriers. Most of the energy released along the chain is utilized for the coupled phosphorylation of ADP to ATP.

For many substrates of importance in respiration, the first intermediate oxidant (or electron carrier) is, as we have seen, NAD^+. The structure of NAD^+ and the mechanism of its reduction to NADH were described in Section 8-3.

NAD^+ oxidizes mainly substrates that are primary or secondary alcohols, or amines. In its turn, NADH is reoxidized to NAD^+ by a flavoprotein, *NADH hydrogenase*, containing FAD or FMN as prosthetic groups (see Section 3-3). In the process, FAD is reduced to FADH.

Several substrates, such as succinate and acyl-CoA complexes, are oxidized directly by dehydrogenase of flavin-protein structure. In this case FAD and not NAD^+ is the first electron acceptor of the chain.

The next electron carrier, capable of oxidizing FADH, is *ubiquinone* or *Coenzyme Q.* Ubiquinone is a benzoquinone derivative with an isoprenoid chain containing 6, 8 or 10 isoprene units. The structure of Coenzyme Q in its reduced and oxidized forms is shown below:

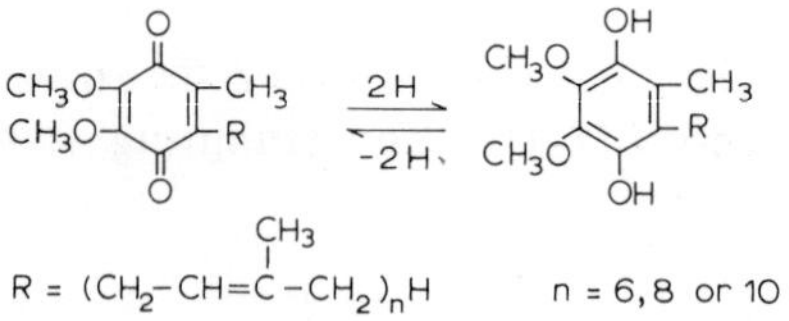

$R = (CH_2{-}CH{=}C(CH_3){-}CH_2)_nH \qquad n = 6, 8 \text{ or } 10$

The ultimate electron carriers of the respiratory chain, the *cytochromes*, are iron-porphyrins (or heme) containing proteins. The heme prosthetic group of cytochromes is similar to that of hemoglobin and myoglobin (Section 5-1-4).

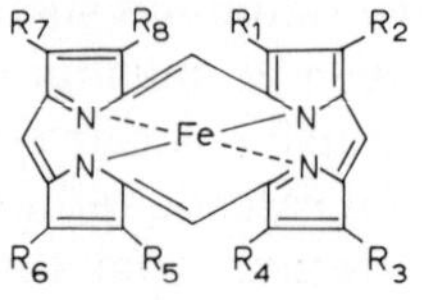

structure of heme

The oxidation–reduction of cytochromes is based on the change of valence of the central iron atom from bivalent to tri-valent and back.

A number of different cytochromes have been identified. They are designated as cytochromes a_1, a_3, b, c, c_1. The sequence of electron transport seems to be $b \rightarrow c_1 \rightarrow a$ or a_3. Cytochromes a and a_3 are the only forms capable of transferring electrons to molecular oxygen and are therefore termed "*cytochrome oxidases*". Cytochrome oxidases (and therefore the whole respiratory chain) are inhibited by cyanide.

The following scheme represents the complete respiratory chain, and indicates the steps that are coupled with the generation of ATP:

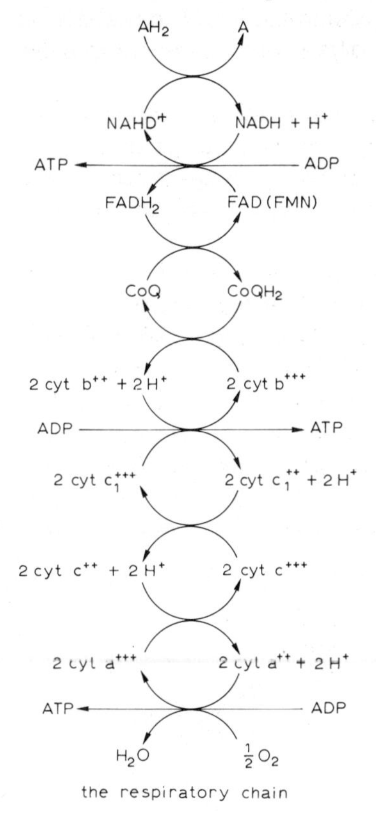

the respiratory chain

18-6. POST-HARVEST RESPIRATION OF FRUITS AND VEGETABLES

18-6-1. *The climacteric pattern.* Harvesting brings to an end the material exchange between the fruit and the rest of the plant. As an independent biological system, the harvested fruit exhibits considerable chemical activity in which respiratory processes play an important role. Under aerobic conditions fruits continue to respire (to absorb O_2 and to expel CO_2), oxidizing their carbohydrate reserves. Most of the energy liberated is evolved as heat and may be measured by calorimetry. Many chemical changes take place and most of these have a direct influence on quality. Some of these changes are: disappearance of astringency and bitterness, changes in acidity, disappearance of chlorophyll and synthesis of some pigments (i.e. change in color), softening of the tissue through breakdown of pectins, development of some aroma constituents, destruction of others, etc. Many of these changes are dependent on or interrelated with respiration.

The rate of respiration of harvested fruits is easily measured by determining the rate of CO_2 emission or oxygen adsorption of fruit in a closed vessel. Today, gas chromatography provides a convenient method of measuring and recording fruit respiration rate continuously.

Typical post-harvest respiration rate curves are shown in Fig. 30. In the case of avocado, respiratory activity falls to a minimum, then rises rapidly to a maximum. Kidd and West, who observed this phenomenon in apples, termed this type of respiratory curve as "*the climacteric*". The climacteric pattern is exhibited by a large number of fruits among which are the apple, pear, peach, apricot, plum, banana, tomato, mango, papaya, avocado, passion fruit, etc. *Ripening*, defined as visible changes in color, texture and flavor, occurs at or shortly after the climacteric peak. In the case of citrus

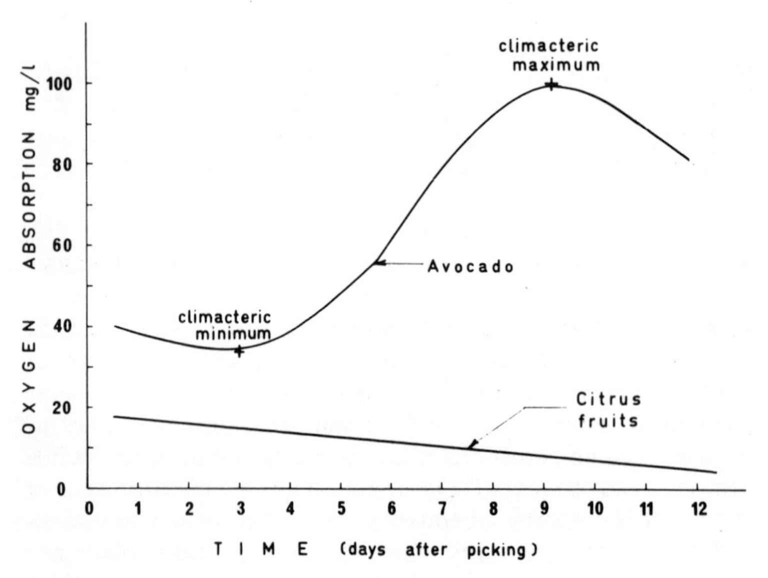

Fig. 30. Typical post-harvest respiration curves of fruits: (a) Avocado; (b) Citrus fruit.

fruit, however, the climacteric pattern is not observed. Biale classified fruits in two groups as climacteric and non-climacteric, the latter group comprising, in addition to citrus fruit, the cherry, fig, grape, strawberry and pineapple. This classification is of practical value but apparently does not reflect a fundamental difference between the two groups. Thus citrus fruit, if picked unripe, exhibits climacteric behaviour. Since citrus is usually picked tree-ripe, many physiologists consider that the non-climacteric character of citrus is due to the fact that the fruit is already at the post-climacteric stage when harvested.

18-6-2. *Effect of ethylene.* The shape of the climacteric curve hints to the autostimulatory character of post-harvest respiration and ripening. The clue to this autostimulation is now accepted to be the biosynthesis of *ethylene* and its effect on post-harvest biochemical activity.

It has been customary to hasten the coloration of green citrus fruits by means of blue-flame kerosene stoves placed directly in the room in which the fruit is stored, the so-called "sweat rooms". For some time it was thought that the change in color is brought about by heat but it has been found that the products of combustion are responsible for the hastening of coloration. Further experiments disclosed that the ripening of the fruit were unsaturated hydrocarbons, the specific gravity of which was approximately equal to that of air. These facts suggested ethylene (C_2H_4), and later experiments proved that this gas, or a mixture of other unsaturated hydrocarbons, could effectively bring about the desired change in color in fruits. The mechanism of this reaction is not yet known. The gases merely stimulate the fruit to renewed life activity. During this period, lasting from five to ten days, the fruit continues to respire by inhaling oxygen and expelling carbon dioxide and, as a result, brings about the discoloration of its green pigment. Even one part by volume of ethylene in five million parts of air can produce the effect. The suitable concentration of the gas, however, depends on the time required for coloration, the location where the process is conducted, and the relative quantity of air present in the space not occupied by the fruit (to provide sufficient oxygen for the respiration of the fruit).

In this connection it is interesting to note that the vapors from moldy lemons inoculated with green mold (*Penicillium digitatum*

Sacc.), cause a very marked increase in the rate of carbon dioxide evolution and greatly accelerate the color development of sound green lemons. Emanations from a single moldy lemon can produce these effects in as many as 500 fruits. Ethylene has been identified as one of the volatile substances given off by fruits during respiration.

Ethylene is present at very low concentration in fruit tissues. Production of ethylene increases sharply with the climacteric rise of respiration rate. The biosynthesis of ethylene is still a controversial subject and a number of pathways have been proposed. One of these, based on methionine as the starting point, is considered at present as the most probable route.

The autocatalytic shape of the climacteric curve may be explained by assuming that ethylene is both a product of ripening and a ripening stimulator. In the light of available experimental evidence, ethylene is regarded as a "*ripening hormone*" which triggers some important reaction in the process. How it does it is not known. It may activate a specific enzyme system or it may enhance enzyme-substrate interaction by increasing the permeability of biological membranes.

18-6-3. *Control of ripening.* The use of ethylene gas for the acceleration of post-harvest ripening of citrus and bananas has been described in the previous paragraph. Long term preservation of fresh fruit often requires the retardation of ripening. This effect can be achieved by *refrigeration*, but low temperature storage can prolong the shelf-life of unprocessed fruit only to a limited extent. The effectiveness of refrigeration alone is often impaired by post-harvest parasites such as fungi. A more efficient method combines refrigeration with *controlled atmosphere* (CA). The rate of respiration of fruits is reduced considerably if the concentration of oxygen in the atmosphere is diminished. In CA storage, the fruit is placed in hermetically sealed refrigerated rooms and the natural atmosphere is replaced by a mixture of gas rich in nitrogen but poor in oxygen. The optimal "formula" of gas is experimentally determined for each set of conditions (fruit variety, initial degree of maturation, desired storage period, temperature). If necessary, the gas in the room is periodically scrubbed to remove excess CO_2. Total removal of oxygen is not desirable. Under completely anaerobic conditions (total depression of respiration), the carbohydrates of the fruit would undergo degradation processes similar to fermentation.

Prolongation of shelf-life through control of respiration rate is also achieved by a novel process named "*hyperbaric storage*", whereby the oxygen tension is reduced not by removal of oxygen as in CA but by lowering the total pressure and storing the fruit under partial vacuum. Another interesting approach makes use of the different permeability of *packaging films* for different gases. The fruit is wrapped in a suitable film. The composition of the atmosphere inside the package is soon altered as the result of respiration. The concentration of oxygen decreases but excessive accumulation of CO_2 is prevented thanks to the high diffusivity of this gas through the film. An equilibrium is reached between consumption of production rate of any gas by the fruit and the rate of diffusion of the same gas through the membrane.

If the film is properly selected, the equilibrium composition of the gas in the package should correspond to a desirable CA formula.

SELECTED BIBLIOGRAPHY TO CHAPTER 18

Axelrod, B., Glycosis, in Metabolic Pathways, 3rd edn., D.M. Greenberg (Ed.), Vol. 1, p. 112, Academic Press, New York, 1967.

Axelrod, B., Other pathways of carbohydrate metabolism, Ibid., p. 272.

Biale, J.B., Postharvest physiology and chemistry, in The Orange, its Physiology and Biochemistry, W.B. Sinclair (Ed.), p. 96, Univ. California Press, Berkeley, 1961.

Florkin, M. and Stotz, E., Comprehensive Biochemistry, Vol. 14 (Biological Oxidations), Elsevier Publishing Co., Amsterdam, 1967.

Krebs, H.A., The intermediary stages in the biological oxidation of carbohydrate, Advances in Enzymol., 3 (1943), 191.

Rhodes, M.J.C., The climacteric and ripening of fruits, in The Biochemistry of Fruits and their Products, A.C. Hulme (Ed.)., Vol. I, p. 521, Academic Press, New York, 1970.

CHAPTER 19

BIOCHEMICAL ACTIVITY OF MICROORGANISMS IN FOODS

19-1. INTRODUCTION

The complexity of foods as biochemical systems is amplified by the presence of microorganisms which contribute their own metabolites and their own catalytic machinery. The mere presence of microorganisms and their chemical activities may constitute a spoilage factor, or on the contrary, microorganisms may be used to bring about a desired change in food characteristics.

The number of types of microorganisms of importance in food technology is very large indeed; Jay lists in this connection twenty-five genera of bacteria, sixteen of molds and nine of yeasts. The detailed study of these microorganisms in relation to food is the task of "*Food Microbiology*", a well developed, richly documented field of modern food science. The purpose of the present chapter is to describe a number of cases of chemical changes in foods, induced by microorganisms.

19-2. ACTION ON CARBOHYDRATES

19-2-1. *Breakdown of polysaccharides and disaccharides.* Many classes of microorganisms can hydrolyze structural polysaccharides, such as pectins and cellulose, by means of extracellular enzymes. Of particular importance in this respect is the spoilage of unprocessed fruits and vegetables by molds. Recently attempts have been made to utilize the cellulolytic activity of molds in solid waste treatment. The idea is to develop a convenient method for the rapid decomposition of urban and industrial waste rich in cellulose (paper, agricultural waste), at the same time producing mycellial mass which can be used as a protein source for animal feeding.

Microbial hydrolysis of disaccharides is sometimes a problem in the sugar industry. Osmophylic yeasts and molds are capable of inverting sucrose in concentrated solutions thereby impairing the crystallization process (see Section 7-4).

19-2-2. *Decomposition of sugars.* Complete decomposition of some carbohydrate to carbon dioxide and water would represent at most a loss of sweetness and caloric value of foods, but would not be a serious spoilage factor. In foods, however, the conditions of pH, lack of air, processing, storage, etc., introduce stress parameters which force the microorganisms to derive their energy in less economic ways. The possibilities are very numerous and so are the terminal products. Following are some examples.

The best studied process of biodegradation of sugars is *alcoholic fermentation. Ethyl alcohol* is the main product of sugar breakdown by yeast under aerobic conditions, provided that the sugar concentration is relatively high and the pH in the range of 3.5 to 5. The initial reactions leading to alcohol follow the Embden–Meyerhof scheme of glycolysis, as discussed in Chapter 18. The last steps involve decarboxylation of pyruvate to acetaldehyde and reduction of the latter to ethanol, coupled with the generation of oxidative power in the form of NAD^+.

$$\begin{matrix} CH_3 \\ | \\ C{=}O \\ | \\ COOH \end{matrix} \xrightarrow[-CO_2]{\text{carboxylase}} \begin{matrix} CHO \\ | \\ CH_3 \end{matrix} \xrightarrow{NADH + H^+ \;\to\; NAD^+} \begin{matrix} CH_2OH \\ | \\ CH_3 \end{matrix}$$

If this pathway is blocked, e.g. by the presence of sulfur dioxide, the reduction step takes place at an earlier stage of glycolysis. One possibility is the reduction of glyceraldehyde to *glycerol.* Varying amounts of glycerol usually accompany ethanol as the final product of alcoholic fermentation. In adequate concentration, glycerol is a normal constituent of wines, essential for the creation of the desirable mouth-feel or "body".

Alcoholic fermentation is the basis of an important industry. As a spoilage problem, it is of significance in apple juice, dry fruit (raisins, dates), juice concentrates and confectionery. Normally, alcoholic fermentation is at its optimum when the pH of the substrate is about 5. However, when the pH is raised into the alkaline region, the course of the yeast fermentation is changed and, under these conditions, glycerol and acetic acid are produced as well as ethyl alcohol and carbon dioxide. In this type of fermentation, acetaldehyde, instead of being reduced to ethanol, undergoes a dismutation in the presence of the enzyme, *aldehyde mutase*, to form alcohol and acetic acid:

$$2CH_3CHO + H_2O \rightarrow C_2H_5OH + CH_3COOH$$

Glycerol, on the other hand, is formed since acetaldehyde is no longer available as an acceptor for NADH, which therefore reduces dihydroxyacetone phosphate to glycerol phosphate. The overall reaction, sometimes called "Neuberg's third form of fermentation", can be expressed as follows:

$$\underset{\text{hexose}}{2C_6H_{12}O_6} + H_2O \rightarrow 2CO_2 + \underset{\text{ethanol}}{C_2H_5OH} + \underset{\text{acetic acid}}{CH_3COOH} + \underset{\text{glycerol}}{2CH_2OH.CHOH.CH_2OH}$$

It has been observed that *Zygosaccharomyces acidifaciens*, a yeast often found in decomposed foods, carries out this type of fermentation.

Another reaction of great importance in foods is lactic acid fermentation. The route to lactic acid was discussed in Chapter 18. A number of microorganisms, notably many streptococci and lactobacilli, ferment sugars quantitatively into lactic acid. These are termed homofermentative microorganisms. The larger class of heterofermentative microorganisms produce, together with lactic acid, a large number of other terminal metabolites, such as acetic acid, ethanol, diacetyl, etc. Lactic acid fermentation is the fundamental process in the manufacture of cultured milk products, cheeses, pickled olives and vegetables, etc. It is also an important cause of spoilage in liquid milk, some fruits (dates, figs) and fruit juices.

In fruit juices, *lactobacilli* carry out a mixed fermentation, the products of which are lactic acid, acetic acid and carbon dioxide.

Production of *acetic acid* is characteristic of Acetobacter sp., which can oxidize ethyl alcohol under aerobic conditions. This fermentation is both an industrial process for the manufacture of vinegar and an important spoilage phenomenon in wines.

A secondary fermentation of importance in the wine industry is the transformation of malic acid to lactic acid or the so-called *malolactic fermentation*:

$$COOHCH_2CHOHCOOH \rightarrow CO_2 + CH_3CHOHCOOH$$

This reaction, carried out by bacteria, is usually desirable. It is an essential process in the ageing of red wines. However, as it causes loss of acidity (one carboxylic group is lost), excessive transformation may be deleterious in wines of low natural acidity.

Stress factors which block the respiratory pathway would cause the accumulation of intermediate metabolites of the Krebs' cycle. In fact, these or their derivatives are often found in spoiled foods. The presence of abnormal quantities of *succinic acid* in eggs and fish has been suggested as an indication of microbial spoilage. At the same time, stress factors are important tools of industrial microbiology. Under certain conditions, selected varieties of *Aspergillus niger* can be made to ferment glucose to *citric acid* in very high yields. This is the principal industrial process for the manufacture of citric acid.

Acetylmethylcarbinol (AMC) and *diacetyl* are two products of carbohydrate degradation, frequently found in spoiled fruit products. Their origin is pyruvic acid and the main microbial agent responsible for their production is Aerobacter. AMC is formed by decarboxylation and condensation of two molecules of pyruvate:

$$2CH_3COCOOH \rightarrow \underset{\text{AMC}}{CH_3{-}CO{-}CH(OH){-}CH_3} + 2CO_2$$

Dehydrogenation of AMC yields diacetyl:

$$CH_3{-}COCH(OH)CH_3 \xrightarrow{-2H} \underset{\text{diacetyl}}{CH_3{-}CO{-}CO{-}CH_3}$$

Diacetyl is sometimes used as a chemical indicator of bacterial spoilage in concentrated orange juice, but it is also one of the components of the characteristic aroma of cultured cream and butter. The interrelation between carbohydrate oxidation and diacetyl formation is illustrated by the fact that addition of citric acid to cream enhances diacetyl production.

Propionic acid, CH_3CH_2COOH, is one of the flavor components produced during the ripening of the famous Swiss *Emmentaler* cheese. It is produced by propionic acid bacteria, mainly by the reduction of lactate.

$$CH_3CH(OH)COOH \xrightarrow{2H} CH_3CH_2COOH + H_2O$$

Other substrates such as various amino acids may also serve as precursors in the production of propionate.

19-2-3. *Bacterial polysaccharides.* The microbial synthesis of polysaccharides from simple sugars is of considerable theoretical interest. In food technology, the production of extracellular bacterial polysaccharide manifests itself as *slime* or *ropiness*. Pasteur showed that the formation of slimes in the sugar industry is due to a "viscous fermentation" induced by bacteria. Bacteria of *Leuconostoc* group, *B. mesentericus* and *B. subtilis* produce extracellular gums such as dextrans and levans (see Chapter 8). Ropiness in bread, milk and pickle brines are examples of spoilage due to the presence of bacterial polysaccharides.

19-3. ACTION ON NITROGENEOUS COMPONENTS

The first and most readily detectable result of biodeterioration of protein rich foods is *putrefaction*. Production of putrid odors is due to the formation of a large variety of volatile odorous substances as the result of microbial breakdown of amino acids, nucleotides and other low molecular weight nitrogeneous compounds. Only after these nutrients have been depleted do the microorganisms carry out proteolysis. Therefore, as a rule, spoiled high protein foods become practically inedible before proteolysis takes place to any considerable extent. Nevertheless, microbial proteolysis is of importance in the ripening of certain cheeses, in the manufacture of soy sauce, etc. (see Section 5-7).

Depending on the microorganism and the conditions, biodeterioration of amino acids may follow different pathways.

(a) *Deamination. Oxidative deamination* produces ammonia and keto acids. The latter may be used by the microorganism as a source of energy with the same end results as carbohydrate decomposition.

$$R-\underset{NH_2}{\underset{|}{CH}}-COOH \xrightarrow[H_2O]{-2H} NH_3 + R-\underset{O}{\underset{\|}{C}}-COOH$$

Reductive deamination is carried out by strict anaerobes and produces ammonia and an unsubstituted carboxylic acid (e.g. a fatty acid)

$$R-\underset{NH_2}{\underset{|}{CH}}-COOH \xrightarrow{2H} RCH_2COOH + NH_3$$

The production of volatile (low molecular weight) fatty acids such as formic, acetic, propionic, butyric and valeric acid is sometimes taken as an indication of spoilage in fish and seafood.

In all cases the ammonia formed is one constituent of the putrid odor.

(b) *Decarboxylation.* Removal of CO_2 is carried out by specific decarboxylases. Anaerobic decarboxylation produces *amines.*

$$RCH(NH_2)COOH \longrightarrow CO_2 + RCH_2NH_2$$

The major cause of putrid odors is often the production of volatile amines. In fish, these are determined as "*volatile bases*", and constitute a widely used chemical index of spoilage. Depending on the parent amino acid, various amines are obtained. The best known are *histamine* from histidine, *cadaverine* from lysine, *putrescine* from ornithine, *tyramine* from tyrosine, etc.

$$\underset{\text{lysine}}{H_2N(CH_2)_4CH(NH_2)COOH} \longrightarrow \underset{\text{cadaverine}}{H_2N(CH_2)_5COOH} + CO_2$$

$$\underset{\text{ornithine}}{H_2N(CH_2)_3CH(NH_2)COOH} \longrightarrow \underset{\text{putrescine}}{H_2N(CH_2)_4COOH} + CO_2$$

histidine → histamine + CO_2 (imidazole ring: $CH{=}CCH_2CH(NH_2)COOH$ with N, NH, CH ring members → $CH{=}C{-}CH_2CH_2NH_2$)

tyrosine → tyramine + CO_2 ($HO{-}C_6H_4{-}CH_2CH(NH_2)COOH \longrightarrow HO{-}C_6H_4{-}CH_2CH_2NH_2 + CO_2$)

(c) *Hydrogen sulfide* is of popular renown as the odorous principle of putrid eggs and other decaying proteinaceous materials. Its source is, of course, cysteine.

(d) Decomposition of the amino acid radical produces sometimes characteristic substances. Typical of the case is the production of *indole* from tryptophane. Escherichia coli, for instance, decomposes tryptophane as follows:

tryptophane (indole ring with $-CH_2-CH(NH_2)COOH$) → indole $+ NH_3 + CH_3COCOOH$

Indole is a useful index of spoilage in fish and seafood.

(e) The putrid odor of spoiled fish is often due to *trimethylamine* (TMA), produced by bacterial reduction of *trimethylamine oxide* (TMAO). TMAO, it will be recalled (Section 18-3), is the typical nitrogenous excretion of marine fish.

19.4. ACTION ON LIPIDS

Microorganisms cannot grow in the absence of water. Pure oils and fats are therefore not vulnerable to microbial attack. The situation is different in fatty foods containing also water. Due to the low interaction between fats and water, even small amounts of water in a large mass of fat can create water activity levels sufficiently high for microbial growth. Thus, mold can grow on butter. The immediate symptom of microbial decomposition of the lipid phase is saponification of triglycerides resulting in a detectable increase in free fatty acid concentration (hydrolytic rancidity). A large number of widespread microorganisms have considerable lipolytic activity.

19-5. PRODUCTION OF TOXINS

The biochemical changes of microbial origin considered so far had detectable effects on food quality factors: flavor, consistency, and sometimes color. It is true that some metabolites such as histamine and tyramine possess characteristic physiological properties but microbially spoiled foods become organoleptically unacceptable long before the concentration of these metabolites reaches dangerous levels.

The situation is entirely different in the case of pathogenic microorganisms. These microorganisms produce highly toxic, specific poisons, termed *toxins.* In certain cases, the toxins are produced in the infected food before its ingestion. The health hazard in this case is associated with ingestion of the toxin itself and it is called "*food intoxication*" (botulism, fungal intoxications). In other cases, the toxin is produced only in the body of the host after foods heavily infested with living cells of the toxin-producing microorganisms are eaten. This type of hazard is called "*food infection*" (viral infections, salmonella infections, etc.).

19-5-1. *Bacterial toxins.* Chemically, bacterial toxins of importance in food-borne poisoning are mostly proteins. In view of their importance, bacterial toxins have been investigated extensively. Unfortunately, the attempts to establish the relationship between the chemical structure of these proteins and their toxicity have not been always successful.

19-5-2. *Fungal toxins (mycotoxins). Ergotism* (intoxication caused by eating cereal grains infested with the fungus *Claviceps purpurea*) is probably the oldest known case of food poisoning. It caused innumerable outbreaks in the Middle Ages, each resulting in thousands of deaths. These epidemics occupied an important place in European folklore, and were often designated as "*ignis sacer*" or the holy fire. The fungal origin of the disease was not discovered until the 18th century. The toxic principles of ergot are numerous alkaloids, most of them derivatives of *lysergic acid.*

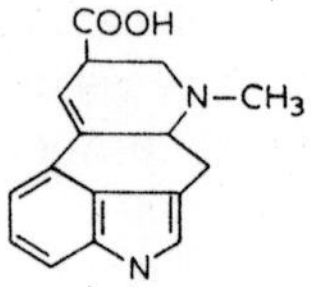

COOH

N–CH_3

N

lysergic acid

In 1960, a mysterious disease (so mysterious that it had to be named the "Turkey X syndrome"), attacked turkey flocks in England and killed as many as 100,000 birds. The cause of death was soon traced back to batches of imported peanut meal used in the feed. The origin of the poison is now known to be a fungus, *Aspergillus flavus.* A. flavus grows on peanut kernels, cottonseeds and other seed crops when these are harvested or stored under conditions of high humidity. Several species produce potent poisons, *aflatoxins.* The structure of one aflatoxin, designated as B_1, is shown below:

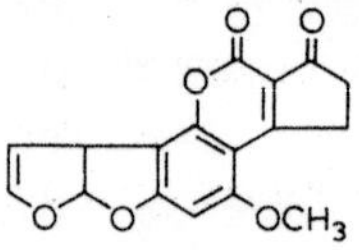

O O

O

O O OCH_3

B_1 aflatoxin

Aflatoxins are not only acute poisons but also potent carcinogens. Ergotism and aflatoxicosis are only two cases of fungal intoxication. Mycotoxins are apparently more widespread than previously thought. The potential presence of toxic substances in mold infested fruits, vegetables and seeds is now viewed with increasing concern.

SELECTED BIBLIOGRAPHY TO CHAPTER 19

Fields, M.L., Richmond, Bonnie S. and Baldwin, Ruth E., Food Quality as Determined by Metabolic Byproducts of Microorganisms, Adv. Food Res., 16 (1968), 161.

Gale, E.F., The Chemical Activities of Bacteria, 3rd edn., University Tutorial Press Ltd., London, 1952.
Jay, J.M., Modern Food Microbiology, Van Nostrand Reinhold Co., New York, 1970.
Rieman, H. (Ed.), Food-borne Infections and Intoxications, Academic Press, New York, 1969.

CHAPTER 20

CHLOROPHYLL AND PHOTOSYNTHESIS

The green thralldom

TANG PEI-SUNG

20-1. INTRODUCTION

The most important of all anabolic processes is *photosynthesis*, namely the conversion of solar energy to chemical energy by green plants. The green pigment of plants, *chlorophyll*, is the major agent capable of absorbing light energy and transmitting it to appropriate carriers, to be used for the synthesis of carbohydrates from carbon dioxide and water. To these carbohydrates we owe the maintenance of life on this planet and it is no wonder that the physiologist Tang Pei-Sung, while writing a book on chlorophyll, expressed our unreserved subordination to this pigment as the "green thralldom".

From the viewpoint of energy flow and overall equation, photosynthesis is the exact reversal of respiration. It involves an *upstream* transport of electrons from water to NAD^+ and finally to organic substrates. Coupled phosphorylation reactions serve to store and release energy much in the same manner as in respiration.

20-2. CHLOROPHYLL

In this section, the chemical structure and properties of chlorophylls will be considered, in connection with the function of chlorophylls as food pigments. The physiological role of chlorophyll in photosynthesis will be considered in the next section.

20-2-1. *Occurrence.* Most of the plants have the chlorophylls in their leaves (before their senescent stage) and in their fruits (before they ripen). A simple microscopic examination shows that chlorophyll is not dissolved in the cells but located in granular structure or plastids, called *chloroplastids* as long as they are green, and *chromoplastids* after they have changed color. The plastids are relatively large, cup-and-plate-like bodies, 3–10 μ in diameter and 1–2 μ in thickness. They are built of smaller particles, *grana* which, because of their size (0.2–2 μ in diameter), lie in the border region of possible detection by the optical microscope. The grana of spinach, for example, are 0.6 μ in diameter and 0.1 μ thick. It is possible by means of an electron microscope to detect even smaller structures, *laminae*, of which the grana are composed. The size range of laminae is only 0.01 μ–0.02 μ. Examined at a magnification of nearly 10^5, the lamella appear as conglomerations of sphere-like sub-units (*quantasomes*) in crystalline array. Within the quantasome lie the molecules of the chlorophylls, apparently bound in some way to the proteins, lipids and liproproteins. Carotenoids, yellow companions of the chlorophylls, and some mineral salts are also found in these positions. The exact orientation of chlorophyll molecules in the laminae is yet unknown. However, there is no doubt that this spongy environment is most appropriate for the specific function of the chlorophylls — the absorption and storage of light energy.

In green leaves of higher plants the chlorophyll content is about 0.1% of the fresh weight. In green algae there appears to be considerable variation in the total content of the green pigment.

20-2-2. *Structure.* Willstraetter and his school were the first to make a thorough study of these pigments and their related compounds and, although they did not establish their final structure, most of their conclusions hold good up to the present. The chlorophylls are always accompanied by yellow pigments, carotene ($C_{40}H_{56}$) and xanthophyll ($C_{40}H_{56}O_2$) (see Chapter 12). In their attempt to elucidate the structural formulae of the chlorophylls, Willstaetter and his collaborators ascertained, in addition to the above, the following important facts: (a) that the molecule of chlorophyll contains one atom of magnesium connected to the rest by a conjugated bond (Willstaetter suggested that Mg acts here in the same manner as in the Grignard reagent and he successfully used this idea when attempting in 1906 a partial synthesis of the tetra-pyrrole ring). It may be pointed out that 100 years before that, De Saussure (1804) showed that Mg is indispensable for the growth of plants; (b) that the chlorophyll molecule consists of four substituted pyrrole rings which are linked together by methine bridges (now conventionally labelled α, β, γ and δ). It is of interest to remark here that the presence of such pyrrole groups in petroleum has been suggested as a proof of the vegetal origin of petroleum; (c) that the tetra-pyrrole structure of chlorophyll is very similar to that of *hemin* (Section 18-5) with the exception that hemin contains iron instead of magnesium; (d) that chlorophyll is the diester formed from a dicarboxylic acid (chlorophyllin), methanol (CH_3OH) and phytyl alcohol ($C_{20}H_{39}OH$), in other words, it is methyl-phytylchlorophyllide; (e) that this magnesium-containing molecule is quite stable in weak alkalis but is easily attacked by even weak acids which cause the cleavage of Mg from the molecule, thereby leaving a subatance with a brownish-olive color (pheophytin).

The structural formula was definitely established by Willstaetter's pupil, Fischer, in 1937. The total synthesis was accomplished only in 1960 by Strell, using a method originally proposed by Fischer and by Woodward from monocyclic pyrroles.

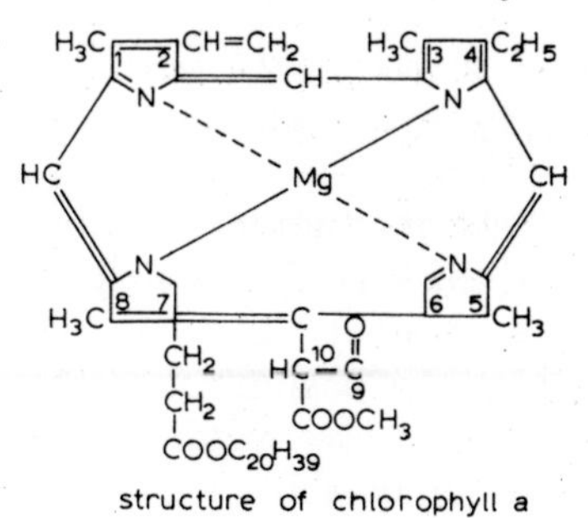

structure of chlorophyll a

Chlorophyll *b* differs from chlorophyll *a* by having an aldehyde group (—CHO) on carbon atom 3 instead of a methyl group.

The special terminology used in chlorophyll chemistry comprises, among others, the following terms:

porphin — the tetra-pyrrole skeleton, consisting of four pyrrole rings connected by four methine bridges;

porphyrin — same skeleton substituted by various groups: methyl, ethyl, vinyl, etc.;

phorbin — the above with the addition of a C_9-C_{10} ring;

chlorin — a phorbin with the C_9-C_{10} ring open;

phorbide — ester of a phorbin;

phaeophorbide — the methyl ester of a phorbin;

phaeophytin — the phytyl ester of a phorbin without the Mg;

phyllin — a derivative of a phorbide or a chlorin with Mg in the center;

chlorophyllin — the dicarboxylic acid:

$$R\left\langle\begin{matrix}COOH\\COOH\end{matrix}\right.$$

methyl chlorophyllide — the methyl ester of the above:

$$R\left\langle\begin{matrix}COOCH_3\\COOH\end{matrix}\right.$$

chlorophyll a or *b* — the above esterified by both methyl and phytyl alcohols:

$$R\left\langle\begin{matrix}COOCH_3\\COOC_{20}H_{39}\end{matrix}\right.$$

Chlorophyll is insoluble in water as long as the phytol moiety is attached to it. However, removal of phytol by hydrolysis yields the water soluble methyl chlorophyllide.

Chlorophyll *a* in acetone solution has a light absorption spectrum with a peak at 663 mμ. In the cell, however, the spectrum is slightly different due to the bound state of the pigment. The in vivo maximum of chlorophyll *a* is at 684 mμ at the beginning of the greening process. It then shifts to 673 mμ. The in vivo maximum of chlorophyll *b* is at 650 mμ. A minor member of the chlorophyll system, accounting for not more than 1/500 of the total chlorophyll, was discovered by Kok in 1956 and named "*P 700*" because of its absorption maximum at 700 mμ. The significance of P 700 in photosynthesis will be considered later.

20-2-3. *Behavior of chlorophyll during food processing.* On the basis of our knowledge of the chlorophyll structure, it is important to summarize the changes that may occur in various food products during processing operations. When it is desirable to retain the full green color of chlorophyll in food products, such as in the case of dehydrated or canned vegetables, the possible changes in the chlorophyll molecules should be kept in mind so as to avoid deterioration of the color.

As mentioned earlier, the action of even a weak acid results in the removal of magnesium from the chlorophyll molecule, and formation of pheophytin, which possesses a brownish-olive color. Although practically every green plant tissue is naturally quite acid, chlorophyll, in situ, is apparently bound to lipoproteins which in some way protect it from the acid action. On the other hand, when heat is applied in food processing, the proteins tend to coagulate and thus the chlorophyll becomes more exposed to the adverse action of the acids. Blanching, therefore, always causes some chlorophyll to be converted into pheophytin. Numerous attempts have been made to protect the chlorophyll by neutralizing a portion of the acid and thus raising the pH of the plant tissue in question by the addition of alkali. However, in most cases, such procedures did not succeed because of marked changes in flavor or in texture. This occurred even when magnesium hydroxide was used as an alkali to reduce acidity in commercially canned green peas.

The enzyme *chlorophyllase* is normally present in plant tissue. The importance of this enzyme in the degradation of chlorophyll during processing has not been demonstrated. Since the action of chlorophyllase is to detach the phytol moiety from the chlorophyll molecule, the net result of its activity would be solubilization of the green pigment in water. The behavior of this enzyme is rather strange: at certain seasons of the year, chlorophyllase remains active at 67° to 75°C, a temperature at which most other enzymes are inactivated, while at other times the same chlorophyllase appears to be relatively inactive. For instance, the chlorophyllase in spinach is practically inactive during the summer months; in the spring, however, the same enzyme will convert chlorophyll in phyllin even after the vegetable has been heated for 20 minutes. Some vegetables, such as peas, string beans and asparagus apparently do not contain the chlorophyllase enzyme at all. That chlorophyllase activity increases with maturity has been demonstrated in many fruits. A number of investigators suggested that hydrolysis of the phytol ester bond may be an essential step in the decomposition of chlorophyll during the ripening of green fruit; however, this view is not supported by experimental evidence.

Some additional phenomena in connection with chlorophyll should be mentioned. Cooking in copper vessels may cause substitution of the magnesium in the chlorophyll molecule by copper. The copper complex of chlorophyll has a deep bluish-green color. Alkaline oxidation will cause *porphyrins* to be formed.

One of the most important mechanisms of chlorophyll degradation in processed vegetables involves oxidation. Oxidative breakdown of the green pigment in frozen or dehydrated vegetable products has been shown to be linked with the enzymic oxidation of lipids (Chapter 15). Thus, the loss of green color in underblanched frozen or dehydrated vegetables could be the indirect result of lipoxygenase-induced lipid oxidation, while discoloration of over-blanched or heat processed vegetables could be attributed, at least partially, to the action of oxidizing free radicals formed from lipids as a result of thermal damage. Although the exact pathway of the degradation reactions is not known, it is assumed that some products of chlorophyll breakdown are responsible for the development of peculiar hay-like odors which accompany the loss of green color. SO_2 is effective in preventing color loss and development off-flavors. It has been claimed that, in this respect, treatment with SO_2 can replace the usual blanching process.

The following scheme summarizes the various changes which may occur in the chlorophylls:

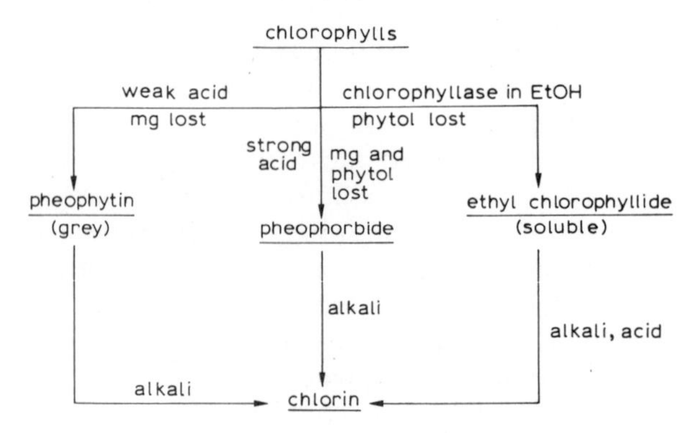

general scheme of chlorophyll breakdown

20-3. PHOTOSYNTHESIS

20-3-1. *Overall equation.* The overall reaction of photosynthesis may be expressed as follows:

$$CO_2 + H_2O \xrightarrow[\text{chlorophyll}]{Nh\nu} (CH_2O) + O_2 \qquad (1)$$

where $Nh\nu$ is the number of quanta required per mol and the term (CH_2O) designates "organic matter", without stating, at the moment, its exact nature.

One could assume from this equation that only one atom of oxygen may be derived from water while the other one may come from CO_2, or that both atoms may be derived from carbon dioxide. The fact is, however, that the only source of oxygen is water, as has been successfully demonstrated by experimentally using water containing the O^{18} isotope. The correct equation will therefore be:

$$CO_2 + 2H_2O^{18} \xrightarrow{Nh\nu} (CH_2O) + H_2O + O_2^{18} \quad (2)$$

At this point it is relevant to mention that some microorganisms and algae are capable of carrying out photosynthesis with compounds other than water. Thus, for instance, green sulfur bacteria are known to oxidize H_2S in the course of carbohydrate biosynthesis:

$$CO_2 + 2H_2S \rightarrow (CH_2O) + H_2O + 2S \quad (3)$$

Therefore, it is possible to represent photosynthesis in a more general way as a hydrogen transfer process:

$$A + DH_2 \xrightarrow{light} AH_2 + D \quad (4)$$

$$2H^+ + H_2D \xrightarrow{light} 2H_2 + D \quad (5)$$

Just as respiration, photosynthesis is an oxidation-reduction reaction. Contrary to respiration, however, the flow of electrons in photosynthesis is in the direction of increasing electronegativity. Light provides the required energy for this "upstream" flow of electrons.

20-3-2. *The light and dark reactions.*

From the above, one can see now that reaction (2) actually consists of two separate reactions, of which the first is engaged in binding hydrogen by some unknown acceptor, while the second reaction is engaged in the reduction of CO_2. We can, therefore, write now:

$$2H_2O + 2A = 2(AH_2) + O_2 \quad (6)$$

$$CO_2 + 2(AH_2) = (CH_2O) + H_2O + 2A \quad (7)$$

The first of these two reactions (*photolysis* of water) requires light energy while the second can proceed in the dark. Thus, photosynthesis can be divided to two distinct phases: the light-induced "upstream" transfer of electrons (the light reaction) and the fixation of CO_2 and its reduction to carbohydrate (the dark reaction).

20-3-3. *Electron transfer in photosynthesis.* We shall now consider in some detail the light reaction. In 1937, Hill succeeded in performing in vitro photolysis of water (equation 6), with isolated chloroplast in the presence of a suitable hydrogen (electron) acceptor A, such as ferric salts, methylene blue or 2 : 6 dichlorophenol indophenol. In 1950, Ochoa and Vishniac performed the Hill reaction, replacing the artificial reducible reagent by $NADP^+$, which was reduced to NADPH. It was speculated that $NADP^+$ could be the electron acceptor of the light reaction in life. In 1954 Arnon and co-workers demonstrated that phosphorylation of ADP to ATP takes place during the light reaction.

Thus the end effects of the light reaction may be summarized as follows:

(a) Water is oxidized. Molecular oxygen is produced.

(b) $NADP^+$ is reduced to NADPH. The latter constitutes the "reductive power", to be used for the reduction of CO_2 to carbohydrate in the course of the dark reaction.

(c) Light energy is transformed to chemical energy and stored as ATP, to be used later for the performance of the endergonic reactions of carbohydrate synthesis.

We must now explain how electrons are "pumped" from the water–oxygen system to the more electronegative system of $NADP^+ \rightarrow NADPH$. This is where light and chlorophyll come into the picture. We have seen that chlorophylls strongly absorb light energy of specific wavelengths. When a molecule is thus "excited" by light energy, it becomes more electronegative, i.e. its tendency to donate electrons increases. Thus, while chlorophyll at ground state cannot reduce $NADP^+$ in a spontaneous manner, photoexcited chlorophyll is sufficiently electronegative to lose an electron to $NADP^+$. The loss of the high energy electron now makes chlorophyll eager to recover its electrons even from relatively electropositive substances such as water. In this manner electrons are "pumped" from water up to $NADP^+$.

The detailed system is more complicated. In green plants, two different pigment systems are believed to exist, operating as two "pumping stations" in series. *Pigment system I* consists mainly of chlorophyll *a* and carotenoids. When photoexcited, it donates high energy electrons to $NADP^+$ through a series of carriers: but even at ground state, pigment system I is too electronegative to replace its electrons at the expense of water. *Pigment system II*, which consists of chlorophylls *a*, *b* and other pigments, absorbs light of shorter wavelength and after excitation donates electrons to pigment system I, again through a series of carriers. At ground state, system II is sufficiently electropositive to accept electrons from water. It is believed that the pigment component of system I (and probably of system II as well) that releases high energy electrons as a result of photo excitation is P 700. Several of the enzymes and carriers of the electron transport chain have been also characterized. Thus *ferredoxin*, an iron-containing protein (not of heme structure), serves as an intermediate carrier in the chain extending from photosystem I to $NADP^+$. Loss and gain of electrons in ferredoxin occur apparently through valence transition of the iron atom. Another carrier, operating between the two pigment systems, is *plastoquinone*, an isoprenoid quinone (c.f. ubiquinone, Section 18-5). Its electron transfer mechanism apparently involves quinone–hydroquinone transition.

$$(H_3C)_2C_6O_2H\text{—}[CH_2\text{—}CH{=}C(CH_3)\text{—}CH_2]_8H$$

Just as in respiration, ATP synthesis is coupled with some of the steps of "downstream" transfer of electrons. A schematic representation of the light reaction is given in Fig. 31 as a hydraulic analogy.

20-3-4. *The dark reaction synthesis of primary carbohydrates.* The main problem confronting scientists during the last hundred years has been the question: which is the primary organic substance formed during the photosynthetic reaction?

It was in 1870 that Von Baeyer suggested that the primary substance created as a result of photosynthesis is *formaldehyde*, HCOH:

$$CO_2 + H_2O \rightarrow HCOH + O_2$$

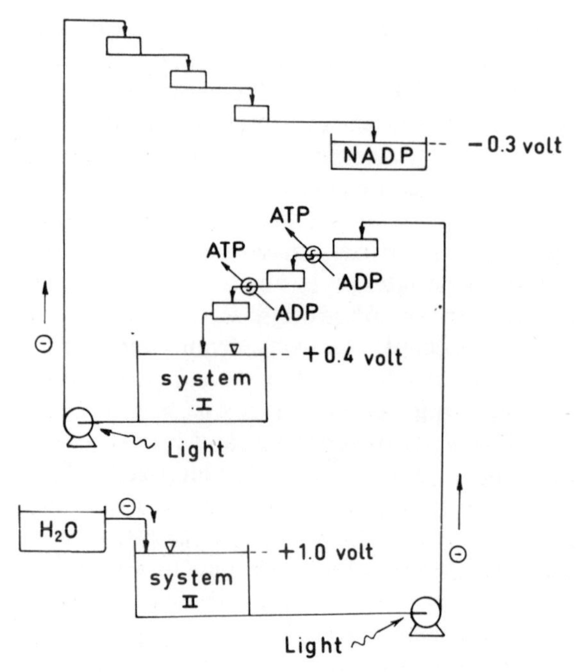

Fig. 31. Path of electron transfer in photosynthesis (hydraulic analogy).

which should, in its turn, give by aldol condensation, a monosaccharide, ($C_6H_{12}O_6$). Willstaetter, who made careful measurements, found that the ratio of CO_2, reduced during the reaction, to oxygen, is one: CO_2/O_2 = I. He therefore supported Von Baeyer's hypothesis *per exclusionem* on the assumption that no other simple organic substance, such as oxalic acid, glycolic acid or formic acid could possibly be formed for, in all these cases, the above ratio of CO_2 to O_2 would differ from unity. Willstaetter assumed, therefore, that photosynthesis is a photochemical reaction proceeding on the surface of the chloroplasts in which both chlorophyll *a* and *b* take part.

Other investigators supporting Von Baeyer's theory assumed it to be a reaction in which Mg took part.

Von Baeyer himself suggested the possibility that formaldehyde formed during the photosynthesis might undergo quick polymerization to form inositol. Von Baeyer thought that by the migration of a hydrogen atom from one carbon to its neighbor a monosaccharide, such as glucose, might be formed.

OH HO OH HO OH OH → OH HO OH HO CHO CH_2OH

inositol carbohydrate

For over half a century, all of these and many other theories on the formation of primary sugars have been investigated. The formaldehyde theory has been finally abandoned, however, since it was never possible to demonstrate the presence of this compound which, in addition, has been found to be toxic to the plant.

In general, the predominant carbohydrate in the leaves of dicotyledone plants is starch, while in monocotyledones glucose is the main carbohydrate. On the other hand, some green algae produce lipids and in the green leaves of wheat a large accumulation of proteins is found right after illumination in the presence of CO_2. Experiments with the green leaves of sunflowers showed the presence of about 50% sucrose, 10% hexoses and 30% starch.

With the advent of modern methods of investigation, and especially with the use of isotopes and chromatography, scientists have succeeded in solving most of the problems connected with this intricate and most valuable process of nature. We owe most of our present-day knowledge regarding the mechanism of CO_2 fixation during photosynthesis to Calvin and his associates. These investigators elaborated a special technique, using both isotope tracers and paper chromatography. For the purpose of these studies, monocellular green algae such as *Chlorella* and *Scenedesmus* were used. The algae were exposed to carbon dioxide, in which the carbon was labelled as the C^{14} isotope, for various periods of illumination. After short periods, which lasted for 5, 30 and 90 seconds and 5 minutes, the algae suspensions were thrown into boiling alcohol which immediately stopped all vital processes, enzymatic and others. The resulting mixtures were now subjected to two-dimensional paper chromatography and the newly formed components, which contained the radioactive carbon, were easily detected by placing the chromatogram on a sheet of photographic paper: wherever compounds containing C^{14} were absorbed onto the chromatogram they affected the photographic paper, leaving spots of varying sizes.

The most important finding during these early investigations was the fact that, after the very first few seconds of illumination, a three-carbon compound, namely *3-phosphoglyceric acid* (PGA) was formed to the extent of 70% of the labelled CO_2 used in the experiments. After 30 seconds the hexoses, glucose and fructose were found and still later sucrose, various amino acids and finally lipids were indicated. All of the carbohydrates were present in the form of esters of phosphoric acid.

The evidence that the primary substance formed was PGA has given no indication as yet as to how it was formed and scientists have been looking in vain for a two-carbon atom compound which might become attached to a molecule of CO_2. It was only after the discovery of two more phosphorylated sugars, a five-carbon carbohydrate (ribulose diphosphate) and the very uncommon seven-carbon carbohydrate (sedoheptulose monophosphate) that light began to dawn about the true mechanism involved.

The first step in this mechanism has been shown to be the reaction between ribulose diphosphate (formed later in the cycle), carbon dioxide and water to form two molecules of PGA. The reaction is catalyzed by diphosphoribulose *carboxy dimutase.*

$$\begin{array}{c} CH_2O-(P) \\ | \\ C{=}O \\ | \\ HCOH \\ | \\ HCOH \\ | \\ CH_2O-(P) \end{array} + CO_2 + H_2O \longrightarrow \begin{array}{c} CH_2O-(P) \\ | \\ HCOH \\ | \\ COOH \\ + \\ COOH \\ | \\ HCOH \\ | \\ CH_2O-(P) \end{array}$$

ribulose-1:3-diphosphate　　　　2 molecules of PGA

The two molecules of PGA now condense into hexose by a series of reactions which are in fact similar to the glycolytic pathway, but in reverse. In order to explain the regeneration of the CO_2 acceptor pentose, Calvin postulated a cyclic process, shown in Fig. 32. It will be noted that the Calvin cycle is very similar to a reversed hexose monophosphate pathway of carbohydrate degradation (Section 18-4).

Experimental evidence gained since the formulation of the Calvin cycle supports the view that this pathway is the major, if not the only, process of photosynthetic hexose production in green plants.

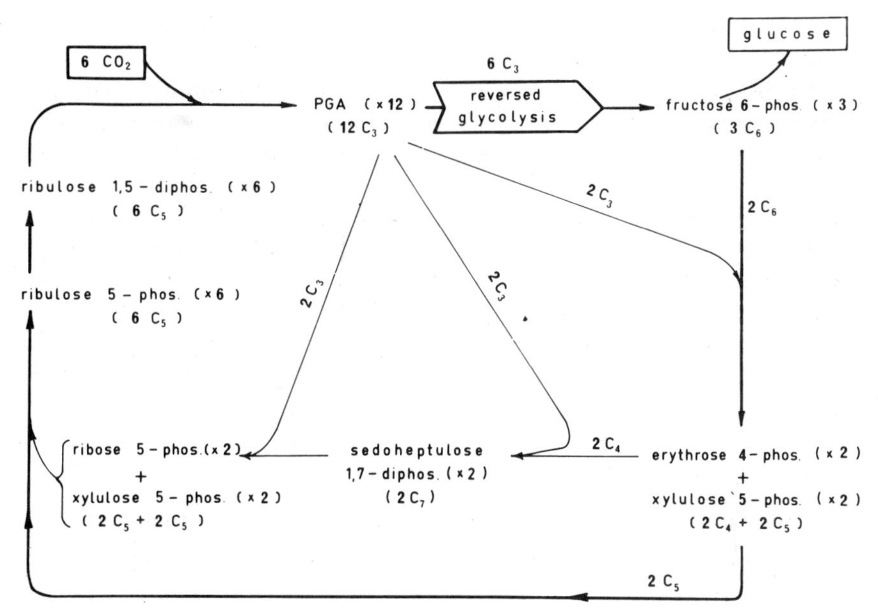

Fig. 32. The Calvin cycle.

20-4. INDUSTRIAL PHOTOSYNTHESIS. "THOUGHT FOR FOOD"

Every year green plants, including the algae in the oceans, bind 150 billion tons of carbon with 25 billion tons of hydrogen to produce organic matter and, at the same time, discharge 400 billion tons of oxygen. But perhaps it is not commonly appreciated that 90% of this enormous chemical "manufacture" is carried out under the surface of sea water by microscopic algae, only 10% of it being carried out by higher plants.

The importance of photosynthesis is not only in the creation of organic matter, viz., food, for all living beings upon the earth, but mainly in storing away at least a part of the enormous solar energy. Our sun sends out 9.10^{22} kcal every second in the form of light energy. Of this energy, only a very insignificant part reaches our planet and of that only a trifle is used by the plants, animals and man. Outside the atmosphere light rays which fall perpendicularly upon an area of one square meter bring along 20 kcal of energy every minute. Only about one-tenth of this energy (2 kcal/min) reaches the Earth and man, himself, merely takes advantage of a few crumbs.

Two crops of corn or potatoes provide 10 million kcal per hectare; raising cattle for milk gives only 1.7 million kcal per hectare, while cattle raised for their meat contribute only 0.25 million kcal; and raising poultry yields 0.5 kcal per hectare. Calculated in another way: 1 kg of corn, eaten as such, provides us with 3500 kcal (100%); the same kg of corn eaten in the form of milk provided by the cow will give us 600 kcal (17%), and eaten as meat, only 350 kcal (10%). It is obvious that protein feeding is a very expensive proposition; in fact, the greatest part of the world's population is fed today mainly on carbohydrate-rich foods poor in proteins.

Theoretically, it is possible to obtain a much higher yield from the photosynthetic process if one cultivates monocellular algae, which grown much faster than higher plants. Some 17,000 varieties of such microscopic algae are known and a few types, such as *Chlorella* and *Scenedesmus*, have been successfully used by scientists in elucidating the intricacies of the photosynthetic process. Such varieties as *Chlorella pyrenoidosa*, *C. vulgaris* and others have been used in recent years for the so-called "Industrial or Controlled Photosynthesis".

Contrary to higher plants, Chlorella contains very little cellulose and, by adjusting the growing media, it is possible to develop such algae with a protein content of up to 50% of the total nutrients or with an augmented content of fats, as the case may be. Algae are very efficient in the use of solar energy in comparison with conventional agriculture. While sugar beets, which are the most efficient converters of solar energy, may produce as much as twenty tons of organic matter per acre per year, controlled photosynthesis, using Chlorella, can yield up to ten times the peak crops obtained today in agriculture. Considering protein alone, the yield differences are even greater. The soybean, the most prolific protein plant, yields up to a ton of protein per acre per year. Chlorella can yield ten times as much plant protein. While Chlorella protein lacks some essential properties necessary for animal feeding, another algae, *Euglena*, produces a protein more like that produced by animals fed on agricultural crops; however, it can produce 100 times more of it than animals do.

Recently, controlled photosynthesis has been very successfully tried in utilizing sewage as a medium for algae growth, and it is claimed that sewage wastes can be reclaimed and re-used an indefinite number of times and, in so doing, can produce unprecedented quantities of food at costs within the economic reach of most societies. In sewage, bacteria decompose organic matter and produce CO_2. The algae utilize this CO_2 for photosynthesis. In turn photosynthesis provides the oxygen necessary for the bio-oxidation of organic matter. Thus a sort of symbiosis is established between bacteria and algae.

The possibility of re-using sewage wastes has been seriously considered by those engaged in outer space travel projects, where the problems of supplying food and oxygen, as well as the disposal of wastes, sewage and carbon dioxide, are very serious.

For centuries, inhabitants of the regions near the lake of Chad in Africa have been using the bluish algae *Spirulina* as a valuable source of food protein. The waters of Lake Chad are of alkaline reaction and therefore can dissolve CO_2 at much higher concentration than neutral media. Spirulina grows successfully in alkaline water and benefits from the large supply of CO_2. Recently, experimental production of Spirulina has been introduced in a number of countries, among them Mexico.

Several pilot plants engaged in industrial photosynthesis have been constructed in various countries. One may now visualize large ponds of water, constructed in any desert, containing the required media exposed to sunshine and, instead of plowing, sowing and otherwise preparing the soil, instead of being dependent on the mercy of rain, these ponds will be seeded through a pump with some cells of Chlorella or other suitable algae. The algae will not suffer the vicissitudes of weather or disease that afflict agricultural crops; they grow so fast so the hazardous interval between planting and harvesting is short. And when the abundant crop is ready, the algae can be collected by centrifugal separation, returning the liquid to another pond and the process started all over again. All this seems very fascinating if it were not for several difficulties which need to be solved by an ingenious chemical engineer:

(1) The growth of Chlorella seems to require sterile conditions, otherwise the danger exists that other parasites among the heterotropic organisms, such as protozoa, yeast and molds, will devour them.

(2) In order to obtain high yields, an optimal content of CO_2 is necessary, which is about 5% in the air. To maintain such a proportion in the open is quite impossible at the present time.

(3) Outside a "closed system" industrial photosynthesis will require an ample supply of CO_2 with its source near the ponds. For each kg of dried Chlorella, 2 kg of CO_2 and 1/12 kg of N are required.

(4) The fact that optimal utilization of solar energy (20%) is attained in dispersed light as against 2% to 3% in full sunlight suggests using an intermittent light source, which means using fluorescent lamps or some other method of darkening the ponds.

(5) The nutritional value of algae is not always satisfactory. Their amino acid compositor is not optimal. Some species, such as Chlorella, have a tough cell-wall and are not easily digested. Algae tend to concentrate products of air and water polluation such as heavy metals and pesticide residues.

(6) Harvesting (separation) and drying of algae are still expensive processes.

These and many other difficulties have still to be overcome. However, large-scale experiments continue and, when they finally succeed, they will disprove the old standing Malthusian Theory* that mankind is doomed to perish due to the ever increasing world population and lack of food.

SELECTED BIBLIOGRAPHY TO CHAPTER 20

Bassham, S.A., Photosynthesis: The Path of Carbon, in Plant Biochemistry, J. Bonner and J.E. Varner (Eds.), p. 875, Academic Press, New York, 1965.

Kok, B., Photosynthesis: The Path of Energy, in Plant Biochemistry, J. Bonner and J.E. Varner (Eds.), p. 904, Academic Press, New York, 1965.

Price, C.A., Molecular Approaches to Plant Physiology, McGraw-Hill Book Co., New York, 1970.

Tang Pei-Sung, Green Thralldom, Allen and Unwin, London, 1949.

Vernon, L.P. and Seely, G.L. (Eds.), The Chlorophylls, Academic Press, New York, 1966.

* Thomas Robert Malthus in 1798.

INDEX